Problem Books in Mathematics

Series Editor

Peter Winkler
Department of Mathematics
Dartmouth College
Hanover, NH 03755
USA
peter.winkler@dartmouth.edu

For further volumes:
http://www.springer.com/series/714

J.W. Neuberger

A Sequence of Problems
on Semigroups

 Springer

J.W. Neuberger
Department of Mathematics
University of North Texas
Denton, TX 76203
USA
jwn@unt.edu

ISSN 0941-3502
ISBN 978-1-4614-3006-3   ISBN 978-1-4614-0430-9 (eBook)
DOI 10.1007/978-1-4614-0430-9
Springer New York Dordrecht Heidelberg London

Mathematics Subject Classification (2010): 46T99, 35K90, 01K12

Printed on acid-free paper

Springer is part of Springer Science+Business Media (www.springer.com)

Contents

Chapter 1
Introduction

Suppose one has an amount of money M to be invested for a year at an annual interest rate I compounded continuously, i.e., total worth of the investment at the end of the year is the limit of what results from compounding $2, 4, 8, 16, \ldots$, in each instance at equal time intervals. During the year, the net worth is continually growing, the rate of earning at each time is proportional to the current value. As the year progresses, the value of the account changes, but the *law* governing earning does not change. This is an example of an autonomous system. One-parameter semigroups in the problems to follow deal with such autonomous systems.

This book consists of a sequence of problems which develop a variety of aspects in the area of one-parameter semigroups of transformations. It is written in the 'Discovery Style' (Moore Method, Texas Method, Inquiry Based Learning, ...) in which definitions and problems, but not proofs, are given. The idea is to give an opportunity for the reader to discover important steps in the development of the subject. The hope is that this style will enable the reader to more quickly arrive at a point of independent research. Paul Halmos, who first informed me of plans for this series of problem books, was an advocate of problem based teaching, as opposed to lecture based teaching.

Problem 1 *Express your opinion on how effective a lecture method might be in teaching someone to ride a bicycle.*

Someone who works through a substantial number of the problems to follow will gain an introduction to a variety of aspects of the subject of one-parameter semigroups of transformations. These problems are not represented as being encyclopedic, but rather seek to give a grounding in various issues connected with semigroup theory.

The problems vary widely in difficulty. Some will be considered easy, some quite difficult and a few pertain to open research questions. In a number of instances references are given in which further background may be found. If a reader finds a problem particularly difficult, they might still consider working hard on it, delaying the pursuit of references for a time. It is hoped that such effort would help a reader to approach a reference from a position of strength.

The final chapter consists of notes on the other chapters. Along with supplying additional information about various problems, the notes give some glimpses of time scales for a research endeavor. They hint at some of the human drama of long-term research quests.

Semigroups, as in Definition 1 give a description of a broad class of semigroups. An underlying theme of this problem sequence is the question as to how semigroups relate to 'time-dependent' differential equations. For a broad class of linear problems, this question was well understood by the 1950s. The corresponding question for a similarly broad class of nonlinear problems started receiving attention in the later 1950s. How the nonlinear theory developed can be pieced together from problems and notes. By now a rather solid foundation for a theory of one-parameter nonlinear semigroups is in place but the the real work is just beginning.

I express my gratitude to Elizabeth Loew of Springer for her help and encouragemeant in preparation of this volume. Out of a legion of teachers, colleagues and students whose influence has helped shape my mathematical thoughts I single out just a few: Alfonso Castro, Bob Dorroh, Jerry Goldstein, David Kendall, Tosio Kato, R.L. Moore and H.S. Wall.

Chapter 2
The Idea of a Semigroup

Problem 2 *Show that if f is a continuous function from R to R so that*

$$f(x) + f(y) = f(x+y) \text{ for all } x, y \in R,$$

then there is $c \in R$ so that

$$f(x) = cx, \ x \in R.$$

Problem 3 *Suppose that g is a continuous function from $[0, \infty)$ to R so that*

$$g(x)g(y) = g(x+y), \ x, y \in [0, \infty). \tag{2.1}$$

Show that either $g(x) = 0$ for all $x \geq 0$, or else there is $b \in R$ so that

$$g(x) = e^{bx}, \ x \geq 0.$$

Problem 4 *Contemplate the possibility of there being more solutions f in Problem 2 or more solutions g in Problem 3 if the word 'continuous' is deleted from the respective statements of these problems. Perhaps don't dwell on the present problem unless the term 'Hamel basis' is familiar.*

Problem 5 *In connection with Problem 4, find and read the second part of Hilbert's Fifth Problem (which is concerned with how algebraic and continuity conditions, taken together, may lead to sufficient differentiability to permit an analysis of a problem).*

Definition 1 *A semigroup on a set X is a function T with domain $[0, \infty)$ and range in the set of all functions from X to X so that*

$$T(0) = I \text{ and } T(t)T(s) = T(t+s), \ t, s \geq 0 \tag{2.2}$$

where $T(t)T(s)$ indicates the composition of the transformations $T(t)$ and $T(s)$. The identity transformation on X is I.

J.W. Neuberger, *A Sequence of Problems on Semigroups*,
Problem Books in Mathematics, DOI 10.1007/978-1-4614-0430-9_2,
© Springer Science+Business Media, LLC 2011

Problem 6 *Find a simple example of a semigroup T.*

Problem 7 *Find an example of a semigroup T that is not so simple.*

Problem 8 *Suppose X is a subset of a Banach space X_0 and F is a function $X \to X_0$ such that if $x \in X$ there is a unique $z : [0, \infty) \to X$ so that*

$$z(0) = x, \; z'(t) = F(z(t)), \; t \geq 0. \qquad (2.3)$$

Denote by T the function with domain $[0, \infty)$ and range in the set of all transformations from $X \to X$ such that if $x \in X$ and $s \geq 0$, then

$$T(s)x = z(s)$$

where z satisfies (2.3). Show that T satisfies (2.2).

Note that for T a semigroup on X and $x \in X$, we write $T(t)x$ instead of the longer $(T(t))(x)$.

One can say, in this case, that F is a generator of T. The term *'generator'* will be used in at least four distinct senses in this collection of problems:

- (i) As the function F for which the solutions z of (2.3) serve to define the semigroup.
- (ii) For a semigroup T on X (if X is a Banach space),

$$F = \{(x, y) \in X^2 : y = \lim_{t \to 0+} \frac{1}{t}(T(t)x - x)\},$$

  the derivative of T at zero. A function *is* a collection of ordered pairs. The above expression for F simply gives the set of all pairs comprising F.
- (iii) Given a semigroup T, the generator is a transformation F from which one can reconstruct T by means of an exponential formula (as we will see later in some problems).
- (iv) In Chapters 17 and 19 there is another notion of generator whose heritage goes back to Gauss and Riemann and then to Sophus Lie. In terms of this kind of generator, a complete theory of nonlinear semigroups was finally established (in about 1992). This will be the subject of a number of problems to follow.

It is often necessary to make understood the sense in which one is using the term 'generator' in a given discussion, but in many cases, a transformation F in items (i),(ii),(iii) is a generator in each of these senses. Often a generator in sense (iv) is related to such an F in a way that would have been familiar to Sophus Lie.

Definition 2 *If X is a topological space one says that a semigroup T on X is strongly continuous if for each $x \in X$ the function g such that*

$$g(t) = T(t)x, \; t \geq 0 \qquad (2.4)$$

is continuous.

Definition 3 *If T is a semigroup on the set X, x ∈ X and g satisfies (2.4), then g is called a trajectory of T.*

Problem 9 *Show that if a semigroup T arises from the setting of Problem 8, then T is strongly continuous.*

Some problems to follow give examples of semigroups, some of which are strongly continuous.

Problem 10 *Suppose $X = [0,1]$ and T is the function with domain $[0,\infty)$ and range the collection of all functions from X to X such that*

$$T(t)x = \frac{x}{1+tx}, \; x \in X, \; t \geq 0.$$

Show that T is a semigroup on X.

Problem 11 *Does there exist a generator for T in Problem 10 in either sense (i) or sense (ii)?*

Problem 12 *Suppose $X = [0,1]$ and T is the function with domain $[0,\infty)$ and range the collection of all functions from X to X so that if $t \geq 0$, then*

$$T(t)x = \begin{cases} 0 & \text{if } t \geq 0 \text{ and } x - t \leq 0, \\ x - t & \text{if } t \geq 0 \text{ and } x - t > 0. \end{cases}$$

Show that T is a semigroup on X.

Problem 13 *Does there exist a generator for T in sense (ii)?*

Definition 4 *If X is a Banach space, and T is a semigroup on X, then T is called linear provided that for each $t \geq 0$, $T(t) \in L(X,X)$, i.e., $T(t)$ is a continuous linear transformation from X to X. Otherwise, T is called nonlinear.*

Definition 5 *If T is a linear semigroup on a Banach space X, then T is said to be continuous if T is continuous as a function from $[0,\infty) \to L(X,X)$, using the operator norm, that is,*

$$\lim_{t \to s} |T(t) - T(s)| = 0 \text{ if } s \geq 0,$$

where if $K \in L(X,X)$, $|K|$ is the least number $M \geq 0$ so that

$$\|Kx\|_X \leq M\|x\|_X, \; x \in X.$$

Problem 14 *Suppose $X = C([-1,1])$, the space of all continuous functions from $[-1,1]$ to R with norm*

$$\|f\|_X = \sup_{t \in [-1,1]} |f(t)|.$$

Denote by F the function such that

$$F(x) = \begin{cases} x, & \text{if } x \le 0, \\ 2x, & \text{if } x > 0. \end{cases}$$

Define T on $[0, \infty)$ so that if $t \ge 0$ and $f \in X$, then

$$(T(t)f)(x) = F(t + F^{-1}(f(x))), \quad x \in [-1, 1], \; t \ge 0. \tag{2.5}$$

Show that T is a semigroup on $C([-1, 1])$.

Problem 15 *Does T in Problem 14 have a generator in the second sense?*

This example is due to G. F. Webb ([73]) and was important in the development of the theory of nonlinear semigroups.

Problem 16 *Suppose $X = \ell_2$, the Hilbert space of all sequences $\{x_k\}_{k=1}^{\infty}$ such that $\Sigma_{k=1}^{\infty} x_k^2 < \infty$ with*

$$\|x\|_X = (\Sigma_{k=1}^{\infty} x_k^2)^{1/2}, \; x \in \ell_2.$$

Define the semigroup T on X so that

$$T(t)(x_1, x_2, x_3, \dots) = (e^{-t}x_1, , e^{-2t}x_2, e^{-3t}x_3, \dots), \; t \ge 0.$$

Show that T is strongly continuous and think about the possibility of a generator in one of the first two senses.

Problem 17 *Show that T in Problem 16, if $t > 0$, then*

$$T(t)^{-1} \text{ exists}$$

but is only densely defined and is nowhere continuous.

Problem 18 *Suppose $X = \ell_2$, the complex Hilbert space of all sequences $\{x_k\}_{k=1}^{\infty}$ such that $\Sigma_{k=1}^{\infty} x_k^2 < \infty$ with*

$$\|x\|_X = (\Sigma_{k=1}^{\infty} |x_k^2|)^{1/2}, \; x \in \ell_2.$$

Define the semigroup T on X so that

$$T(t)(x_1, x_2, x_3, \dots) = (e^{-it}x_1, e^{-2it}x_2, e^{-3it}x_3, \dots), \; t \ge 0.$$

Show that T is strongly continuous and think about the possibility of a generator in one of the first two senses.

Problem 19 *Show that T in Problem 18 has the property that*

$$T(t)^{-1} \text{ exists}$$

and is continuous with domain all of X, for all $t \in R$. This semigroup T is actually a group. It is fundamental to the study of the Schrödinger equation of quantum mechanics.

Problem 20 *Show that neither of the semigroups in Problems 16 and 18 is continuous.*

Problem 21 *Suppose $X = [0, \infty)$. Find solutions u to*

$$u(0) = x \in X, \ u' = u^2. \tag{2.6}$$

Show that if $x \in X, x > 0$, then no solution u (solution has 'blowup') to (2.6) exists on all of X (one might solve (2.6) in this case in closed form in order to observe this).

Problem 22 *Following Problem 21 consider trying to make a semigroup that arises from this problem in the manner of (looking ahead) Problem 315. Articulate any difficulties you encounter. Consider possible modifications of Definition 1 that might enable Equation 2.6 to be included in a broad study of semigroup theory. Eventually see Chapter 19.*

Problem 23 *Suppose $X = [0, \infty)$. Find solutions u to*

$$u(0) = x \geq 0 \in X, \ u' = -u^2. \tag{2.7}$$

Show that (2.7) has a solution on all of $[0, \infty)$. Exhibit a semigroup generated by (2.7).

Chapter 3
Translation Semigroups

Problem 24 *Suppose* $X = C([0, \infty))$, *the Banach space of all bounded continuous functions from* $[0, \infty)$ *to* R *with norm*

$$\|f\|_X = \sup_{t \geq 0} |f(t)|.$$

Define the semigroup T *on* X *by*

$$(T(t)f)(x) = f(x + t), \quad x, t \geq 0, \ f \in C([0, \infty)). \tag{3.1}$$

Is T *strongly continuous?*

Problem 25 *Same as Problem 24 except that* X *is the set of all functions which are bounded and uniformly continuous from* $[0, \infty)$ *to* R. *Is the resulting semigroup* T *strongly continuous?*

Problem 26 *Is either of the semigroups in Problems 24 and 25 continuous (see Definition 5)?*

Problem 27 *Suppose* $f, g : [0, \infty) \to R$ *are continuous and if* $t \geq 0$, *then*

$$g(t) = \lim_{h \to 0+} \frac{1}{t}(f(t + h) - f(t)) \text{ for all } t \geq 0.$$

Show that f *is differentiable on* $[0, \infty)$.

Problem 28 *Denote by* X *the Banach space, under sup norm, of all bounded continuous functions from* $[0, \infty)$ *to* R. *Define* T *as in (3.1) and define*

$$A = \{(f, g) \in X :$$

$$g(x) = \lim_{t \to 0+} \frac{1}{t}((T(t)f)(x) - f(x)), \ f \in X = [0, 1], \ x \geq 0\},$$

the generator for T *in the second sense. Show that*

$$Af = f' \in X \text{ if } f \in D(A), \text{ the domain of } A.$$

J.W. Neuberger, *A Sequence of Problems on Semigroups*,
Problem Books in Mathematics, DOI 10.1007/978-1-4614-0430-9_3,
© Springer Science+Business Media, LLC 2011

Problem 29 *For A as in Problem 28, show that if $g \in X$ and $\lambda > 0$, then there is one and only one $f \in X$ such that f is in the domain of A and*

$$f - \lambda A f = g.$$

Problem 30 *For A, g as in Problem 29, show that f in that problem is given by*

$$f(x) = \frac{1}{\lambda} \int_0^\infty e^{-r/\lambda} g(r + x)\, dr, \; x \geq 0. \tag{3.2}$$

In a sense, f is a Laplace transform of g.

Problem 31 *Take A as in Problem 28. Show that if $\lambda \geq 0$, then*

$$(I - \lambda A)^{-1}$$

exists, is a member of $L(X, X)$ and

$$|(I - \lambda A)^{-1}| \leq 1.$$

For Problems 32 through 36, suppose that A is as in Problem 28, $x \geq 0$, $\lambda > 0$ and $f \in X$ where X is the Banach space (under sup norm) of all bounded, real-valued uniformly continuous functions on $[0, \infty)$.

Problem 32 *Show that*

$$((I - (\lambda/2)A)^{-2} f)(x)$$
$$= (\frac{2}{\lambda})^2 \int_0^\infty \int_0^\infty \exp(-(2/\lambda)(s_1 + s_2)) f(s_1 + s_2 + x)\, ds_1\, ds_2. \tag{3.3}$$

Problem 33 *Convert (3.3) to a single integral (rotate coordinates 45 degrees) to obtain*

$$((I - (\lambda/2)A)^{-2} f)(x) = (\frac{1}{\lambda})^2 \int_0^\infty \exp(-(2s/\lambda)) s f(s + x)\, ds. \tag{3.4}$$

Problem 34 *Suppose n is a positive integer. Show that*

$$((I - (\lambda/n)A)^{-n} f)(x)$$
$$= (\frac{n}{\lambda})^n \int_0^\infty \exp(-(ns/\lambda))(s^{n-1}/(n-1)!) f(s + x)\, ds. \tag{3.5}$$

Problem 35 *Show that (3.5) may be rewritten, using a Stieltjes integral, as*

$$((I - (\lambda/n)A)^{-n} f)(x) = \int_0^\infty f(s + x)\, d\phi_{\lambda,n}(s) = \int_0^\infty f(s + x) \phi'_{\lambda,n}\, ds, \tag{3.6}$$

where

$$\phi_{\lambda,n}(s) = 1 - \sum_{k=0}^{n-1} \exp(-ns/\lambda) \frac{(ns/\lambda)^k}{k!}, \quad s \geq 0. \tag{3.7}$$

Problem 36 *Show that if $\lambda > 0$, $x \geq 0$, then*

$$\lim_{n \to \infty} ((I - (\lambda/n)A)^{-n} f)(x) = f(x + \lambda). \tag{3.8}$$

Some advice may be in order here. Equation 3.7 gives a well-known (at least to probabilists) Poisson distribution.

Problem 37 *Using the notation of Problem 35, show that*

$$\lim_{n \to \infty} \phi_{\lambda,n}(s) = \alpha_\lambda(s)$$

where α_λ is defined by

$$\alpha_\lambda(s) = \begin{cases} 0 & \text{if } 0 \leq s < \lambda, \\ 1/2 & \text{if } s = \lambda, \\ 1 & \text{if } s > \lambda. \end{cases}$$

Before working Problem 36, one might pause to review some probability theory concerning the law of large numbers (central limit theorem). The problem *can* be worked from the beginning by careful examination of the terms in (3.5) or (3.7), first noting which term is maximum and how the other terms 'fall off' as one proceeds from the mean in either direction.

The theory of semigroups has a rich connection with probability theory. It is rather widely known that semigroups have many applications to probability theory. What is not so well appreciated is that probability theory has a number of significant *applications* to semigroup theory (cf. [16]).

Chapter 4
Linear Continuous Semigroups

In this chapter we suppose that X is a Banach space with norm $\|\cdot\|$ and T is a semigroup on X such that

$$T(t) \in L(X, X), t \geq 0$$

where $L(X, X)$ represents the ring of all continuous linear transformations from X to X. Suppose also that T is continuous in the sense of Definition 5.

If $C \in L(X, X)$ denote by $|C|$ the norm of C, that is, the smallest non-negative number such that

$$\|Cx\| \leq |C| \|x\|, \ x \in X.$$

We will eventually see that the difference between a semigroup being continuous and its being merely strongly continuous is an important distinction. Continuous linear semigroups arise from what are essentially ordinary differential equations (even though the underlying space may be infinite dimensional). Strongly continuous semigroups that are not continuous can pertain to partial differential equations — a great difference! Note that 'strongly continuous' in Definition 2 is a notion weaker than the notion of 'continuous' in Definition 5. The terminology is an historical accident and is somewhat unfortunate. In Chapter 5 problems will deal with strongly continuous semigroups. Some of the problems in this chapter partially prepare one for problems in Chapter 5.

Problem 38 *If $t, s \geq 0$ and T is a continuous linear semigroup, show that*

$$(T(t) - I) \int_0^s T(r) \, dr = (T(s) - I) \int_0^t T(r) \, dr. \tag{4.1}$$

Problem 39 *If $s > 0$ and T is as in Problem 38, show that*

$$\lim_{t \to 0+} \frac{1}{t}(T(t) - I)\frac{1}{s} \int_0^s T(r) \, dr$$

exists in $L(X, X)$.

J.W. Neuberger, *A Sequence of Problems on Semigroups*,
Problem Books in Mathematics, DOI 10.1007/978-1-4614-0430-9_4,
© Springer Science+Business Media, LLC 2011

Problem 40   *Suppose that T is as in Problem 39. Show that*

$$\lim_{s \to 0+} \frac{1}{s} \int_0^s T(r)dr = I.$$

Problem 41   *Suppose that T is as in Problem 39. Show that there is $B \in L(X, X)$ such that*

$$\lim_{t \to 0+} \frac{1}{t}(T(t) - I) = B.$$

Note that B is a generator of T in the second sense.

Problem 42   *Suppose that T is as in Problem 39. Show that*

$$T'(t) = BT(t), \ t \geq 0$$

and

$$T(t) = I + \int_0^t BT(r) \, dr, \ t \geq 0.$$

Problem 43   *Suppose that $\lambda \in R$ is an eigenvalue of B as in Problem 42, i.e., there is $g \in X$ not equal to zero such that*

$$Bg = \lambda g.$$

Show that

$$T(t)g = \exp(t\lambda)g, \ t \geq 0.$$

Problem 44   *Suppose that T is as in Problem 39, n is a positive integer and that $X = R^n$. Suppose also that $\lambda_1, \lambda_2, \ldots, \lambda_n$ is a collection of distinct eigenvalues of B and that $x_1, x_2, \ldots, x_n$ is a corresponding sequence of eigenvectors with*

$$Bx_k = \lambda_k x, \ k = 1, 2, \ldots, n.$$

Show that if $x \in X$ and

$$x = c_1 x_1 + \cdots + c_n x_n,$$

then

$$T(t)x = c_1 \exp(\lambda_1 t)x_1 + \cdots + c_n \exp(\lambda_n t)x_n, t \in R.$$

Problem 45   *Give an appropriate generalization of Problem 44 to the case in which X is a finite-dimensional vector space over the complex numbers and some eigenvalue may have multiplicity greater than one (recall the Jordan normal form theorem).*

Some of the next problems may help to give an alternative for determining T from a generator.

Problem 46 *Suppose X is a Banach space, $B \in L(X, X)$, and $f_0, f_1, f_2, \ldots$ is a sequence of continuous functions from $[0, \infty) \to X$ such that*

$$f_n(t) = I + \int_0^t B f_{n-1}(r) \, dr, \ t \geq 0, n = 1, 2, \ldots.$$

Show that

$$\|f_{n+1}(t) - f_n(t)\| \leq |B| \int_0^t \|f_n(r) - f_{n-1}(r)\| \, dr$$

for $t \geq 0, n = 1, 2, \ldots.$

Problem 47 *Using the notation of Problem 46, show that if $c > 0$, then there is $K > 0$ such that*

$$\|f_{n+1}(t) - f_n(t)\| \leq K \frac{|B|^n}{n!}, \ t \in [0, c], n = 1, 2, \ldots.$$

Problem 48 *Show that $\{f_n\}_{n=1}^\infty$ is uniformly Cauchy on $[0, c]$ for all $c > 0$. Denote by f the function with domain $[0, \infty)$ such that $\{f_n\}_{n=1}^\infty$ converges uniformly in each $[0, c]$ for all $c > 0$. Show that*

$$f(t) = I + \int_0^t B f(r) \, dr, \ t \geq 0.$$

(Method of successive approximations, Picard's method.)

Problem 49 *Show that*

$$T(t) \ = \ e^{tB} \ = \ \Sigma_{n=0}^\infty \frac{(tB)^n}{n!}, \ t \geq 0$$

where the series converges in the norm of $L(X, X)$.

Problem 50 *Show that if $\lambda \geq 0$, then*

$$T(\lambda) = \lim_{n \to \infty} (I + \frac{\lambda}{n} B)^n. \tag{4.2}$$

Problem 51 *Suppose $C \in L(X, X)$. Show that if $\lambda \geq 0$, then*

$$\lim_{n \to \infty} (I + \frac{\lambda}{n} C)^n \tag{4.3}$$

exists and if S with domain $[0, \infty)$ is such that $S(\lambda)$ is this limit for all $\lambda \geq 0$, then S is a continuous semigroup.

Problem 52 *Suppose $C \in L(X, X)$. Show that if $t > 0$ and $t|C| < 1$, then*

$$(I - tC)^{-1}$$

exists and is in $L(X, X)$.

Problem 53 *For S as in Problem 51 show that*

$$S(\lambda) = \lim_{n \to \infty} (I - \frac{\lambda}{n} C)^{-n}.$$

Chapter 5
Strongly Continuous Linear Semigroups

Suppose X is a Banach space. Here are some problems concerning the class of linear semigroups T which are strongly continuous and have the property that $|T(t)| \leq 1$, $t \geq 0$, that is, T is a strongly continuous semigroup of contractions (T is also called a nonexpansive semigroup). The contraction property makes our investigation a little easier but the general case of strongly continuous linear semigroups is actually an application of the contraction case. First we use a generator of T in the second sense:

$$A = \{(x,y) \in X^2 : y = \lim_{t \to 0+} \frac{1}{t}(T(t)x - x)\}. \tag{5.1}$$

For each $\lambda > 0$ denote by I_λ the transformation so that

$$I_\lambda x = \frac{1}{\lambda} \int_0^\infty e^{-r/\lambda} T(r)x \, dr, \ x \in X. \tag{5.2}$$

Problem 54 *Show that if $\lambda > 0$, then $|I_\lambda| \leq 1$.*

Problem 55 *Show that if $x \in X$, then*

$$\lim_{\lambda \to 0+} I_\lambda x = x.$$

Problem 56 *Show that if $x \in X$, then $I_\lambda x \in D(A)$, the domain of A.*

Problem 57 *Show that if $\lambda > 0$ and $x \in X$, then*

$$(I - \lambda A)I_\lambda x = x,$$

that is, $I - \lambda A$ is a left inverse of I_λ.

Problem 58 *Show that if $x \in D(A)$, then*

$$I_\lambda (I - \lambda A)x = x,$$

that is, $(I - \lambda A)$ is also a right inverse of I_λ.

J.W. Neuberger, *A Sequence of Problems on Semigroups*,
Problem Books in Mathematics, DOI 10.1007/978-1-4614-0430-9_5,
© Springer Science+Business Media, LLC 2011

Some help with this problem follows.

Problem 59 *Suppose $x \in D(A)$ and define $h : [0,\infty) \to X$ as*

$$h(t) = T(t)x, \ t \geq 0. \tag{5.3}$$

Show that the right derivative h^+ of h exists in all of $[0,\infty)$ and $h^+(t) = T(t)Ax$, $t \geq 0$. Show also that h^+ is continuous.

Problem 60 *Show that for h as in Problem 59, h' exists on $[0,\infty)$.*

Problem 61 *Show that h in Problem 59 satisfies*

$$h'(t) = T(t)Ax, \ t \geq 0,$$

and

$$h'(t) = Ah(t), \ t \geq 0 \tag{5.4}$$

provided that $x \in D(A)$, the domain of A.

Problem 62 *Suppose that $x \in X$ but x is not in $D(A)$. Show that there is a sequence $\{x_n\}_{n=1}^{\infty}$ of members of $D(A)$, converging to x so that if $c > 0$, then*

$$\{T(\cdot)x_n\}_{n=1}^{\infty}$$

converges uniformly to

$$T(\cdot)x \tag{5.5}$$

on $[0,c]$.

Definition 6 *The expression in (5.5) is called a generalized solution of (5.4).*

Definition 7 *Suppose G is a transformation from a subset of X into X. The statement that G is closed means that*

$$\{(x,Gx) : x \in D(G)\} \text{ is a closed subset of } X \times X.$$

Problem 63 *Show that if $\lambda \geq 0$, then $(I - \lambda A)^{-1}$ is closed and also that A is closed.*

Problem 64 *Show that $A \in L(X,X)$ if and only if $D(A) = X$. (Use the closed graph theorem.)*

Problem 65 *Suppose $\lambda > 0$, $x \in X$ and m,n are positive integers. Show that*

$$(I_{\lambda/n})^m x =$$

$$\left(\frac{n}{\lambda}\right)^m \int_0^{\infty} \cdots \int_0^{\infty} e^{-(n/\lambda)(s_m+\cdots+s_1)} T(s_m + \cdots + s_1)x \ ds_m \cdots ds_1$$

$$= \left(\frac{n}{\lambda}\right)^m \int_0^{\infty} e^{-sn/\lambda} \frac{s^{m-1}}{(m-1)!} T(s)x \ ds.$$

In particular,

$$(I_{\lambda/n})^n x = \int_0^\infty d\phi_{\lambda,n}\, T(\cdot)x$$

where

$$\phi_{\lambda,n}(s) = 1 - \Sigma_{k=0}^{n-1} e^{-ns/\lambda}\frac{(ns/\lambda)^k}{k!},\ s \geq 0. \qquad (5.6)$$

This is the same distribution as in (3.7).

Problem 66 *Show that if $x \in X$ and $\lambda \geq 0$, then*

$$\lim_{n\to\infty}(I - \frac{\lambda}{n}A)^{-n}x = T(\lambda)x. \qquad (5.7)$$

Problem 66 is, essentially, half of the famous theorem of Hille–Yosida for linear strongly continuous semigroups of contractions: Given one of these semigroups, define its generator (in the second sense) and reconstruct the semigroup from its generator by means of an exponential formula) (5.7).

The problems that follow in this sequence will make the other half of the Hille–Yosida theorem in the present case: Suppose A is a closed linear transformation with dense domain in X, with the property that $(I - \lambda A)^{-1}$ exists, with domain all of X and $|(I - \lambda A)^{-1}| \leq 1, \lambda \geq 0$.

Problem 67 *Show that*

$$A(I - \lambda A)^{-1} = \frac{1}{\lambda}((I - \lambda A)^{-1} - I). \qquad (5.8)$$

Denote the expression in (5.8) by A_λ and call it the Yosida approximation to A at λ.

We want to construct a semigroup T which has A as its generator.

Problem 68 *Show that if $x \in D(A)$, then*

$$\lim_{\lambda\to 0+} A_\lambda x = Ax.$$

If $\lambda > 0$ denote by T_λ the semigroup with generator A_λ, $t \geq 0$, that is,

$$T_\lambda(t) = e^{tA_\lambda}.$$

Problem 69 *Show that*

$$|T_\lambda(t)| \leq 1,\ t \geq 0.$$

Problem 70 *Show that if $\lambda > 0$, then T_λ is a continuous linear semigroup.*

Problem 71 *Show that if $\alpha, \beta > 0$, then*

$$T_\alpha(t)(I - \beta A)^{-1} = (I - \beta A)^{-1}T_\alpha(t),\ t \geq 0.$$

Problem 72 *Show that if $\alpha, \beta > 0$, then*

$$T_\alpha(t)T_\beta(s) = T_\beta(s)T_\alpha(t), \ t, s \geq 0.$$

Problem 73 *Suppose that $n \in Z^+$ and each of*

$$\{C_k\}_{k=1}^n, \{D_k\}_{k=1}^n \in L(X, X),$$

$x \in X$ and

$$|C_k|, |D_k| \leq 1, \ k = 1, \ldots, n.$$

Find an inequality for

$$\|C_1 C_2 \cdots C_n x - D_1 D_2 \cdots D_n x\|.$$

Problem 74 *Suppose $\alpha, \beta > 0$. Show that*

$$\|T_\alpha(t)x - T_\beta(t)x\| \ \leq \ t\|A_\alpha x - A_\beta x\|, \ x \in X, t \geq 0.$$

Problem 75 *Show that there is a strongly continuous semigroup T of contractions such that*

$$T(t)x \ = \ \lim_{\lambda \to 0+} T_\lambda(t)x, \ x \in X, \ t \geq 0.$$

Problem 76 *Show that*

$$A \ = \ \{(x, y) \in X^2 : y = \lim_{t \to 0+} \frac{1}{t}(T(t)x - x)\} \tag{5.9}$$

where T is as in Problem 75.

Problem 77 *Show that A of the preceding problem is a generator of T in each of the first three senses.*

Now suppose that T is a linear strongly continuous semigroup which is not a semigroup of contractions.

Problem 78 *Show that there exists $M > 0$ so that*

$$|T(t)| \ \leq \ M, \ t \in [0, 1].$$

(Recall the theorem of uniform boundedness.)

Problem 79 *Show that for M, T as in Problem 78 and $w = \ln(M)$, it is true that*

$$|T(t)| \ \leq \ Me^{wt}, \ t \geq 0.$$

Problem 80 *For M, T, w as in Problem 79, define S by*

$$S(t) \ = \ e^{-wt}T(t), \ t \geq 0.$$

Show that S is a strongly continuous semigroup and that

$$|S(t)| \leq M, \, t \geq 0.$$

Problem 81 *For M, T, w, S as in Problem 80, define a norm $\|\cdot\|'$ by*

$$\|x\|' = \sup_{t \geq 0} \|S(t)x\|, \, x \in X.$$

Show that the norm $\|\cdot\|'$ is equivalent to $\|\cdot\|$ in the sense that there are k, $K > 0$ such that

$$k\|x\|' \leq \|x\| \leq K\|x\|', \, x \in X.$$

Show that S is a semigroup of contractions (i.e., nonexpansive) under the norm $\|\cdot\|'$.

Problem 82 *Make an analysis of the semigroup T by means of a study of S, using the fact that S, under the norm $\|\cdot\|'$, is a semigroup of the type studied in Problems 56–76.*

In (5.7) we have a generalization of formula

$$e^x = \lim_{n \to \infty} (1 - \frac{x}{n})^{-n}, \, x \in R.$$

Problem 83 *Discuss the possibility of using*

$$\lim_{n \to \infty} (I + \frac{\lambda}{n} A)^n x, \, x \in X$$

in place of one suggested by (5.7).

Problem 84 *Articulate why the second alternative below is more likely to be true than the first alternative:*

$$T(t)x = \lim_{n \to \infty} (I + \frac{t}{n} A)^n x \in X, t \geq 0,$$

$$T(t)x = \lim_{n \to \infty} (I - \frac{t}{n} A)^{-n}, x \in X, t \geq 0.$$

Chapter 6
An Application to the Heat Equation

Denote by H the subspace of $L_2([0,1])$ consisting of all $g \in L_2([0,1])$ for which there is $f \in L_2([0,1])$ such that for some $c \in R$,

$$g(t) = c + \int_0^t f, \ t \in [0,1]. \tag{6.1}$$

In this case f is denoted as g', and is considered to be a generalized derivative of g. Denote by H the vector space of all functions g as in (6.1) with

$$\|g\|_H^2 = \|g\|_{L_2([0,1])}^2 + \|g'\|_{L_2([0,1])}^2.$$

Problem 85 *Show that H is a Hilbert space.*

H will also be denoted by $H^{1,2}([0,1])$ and is called a Sobolev space. An alternate, but equivalent definition will be given later in this problem sequence.

Problem 86 *Suppose that*

$$f(x) = \begin{cases} 0 & \text{if } 0 \le x < 1/2, \\ 1/2 & \text{if } x = 1/2, \\ 1 & \text{if } 1/2 < x \le 1, \end{cases}$$

and g is as in (6.1). Critique the assertion that $g' = f$. Show that all members of H are continuous.

Problem 87 *Suppose that H is as above and H_0 is the subspace of H so that*

$$H_0 = \{f \in H : f(0) = 0 = f(1)\}$$

and that

$$A = \{(f, f'') : f \in H_0, f' \in H\}. \tag{6.2}$$

Show that A is the generator of a strongly continuous semigroup T on H_0.

J.W. Neuberger, *A Sequence of Problems on Semigroups*,
Problem Books in Mathematics, DOI 10.1007/978-1-4614-0430-9_6,
© Springer Science+Business Media, LLC 2011

Problem 88 *For the setting in Problem 87, show that if* $u : [0, \infty) \times [0, 1] \to R$ *is defined by*

$$u(t, x) = (T(t)f)(x), \ t \geq 0, \ x \in [0, 1],$$

then

$$u(0, x) = f(x), u_1(t, x) = u_{2,2}(t, x), u(t, 0) = 0 = u(t, 1), x \in [0, 1], t \geq 0$$
$$(6.3)$$

where $u_1(t, x)$ *is the partial derivative of* u *of the first order with respect to the first argument of* u *at the point* (t, x) *and* $u_{2,2}(t, x)$ *is the partial derivative of* u *of the second order in the second argument of* u. *The partial derivatives are taken in the generalized sense above. This is the famous heat equation.*

The next two problems in this chapter deal with numerical problems in approximating semigroups. There are two reasons for these problems. The first is to introduce some useful numerical ideas and the second is to illustrate the First Law of Numerical Analysis:

> **'Numerical difficulties and analytical difficulties always come in pairs.'**

This is illustrated with the heat equation in Problem 88.

Problem 89 *Suppose* $n, N \in Z^+$ *and*

$$u^{0,k} = f(k/n), \ k = 1, \ldots, n - 1, \ u^{0,0} = 0 = u^{0,n}, \qquad (6.4)$$

where f *is the function given as initial data in* (6.3), *being in this case continuous.*

Given $w > 0$ *and an integer* N, *write a computer program to calculate*

$$u^{j,k} , k = 1, \ldots, n - 1, \ j = 1, 2, \ldots, N \qquad (6.5)$$

such that

$$u^{j,0}, \ u^{j,n} = 0, \ j = 1, \ldots, N \qquad (6.6)$$

and

$$\frac{u^{j,k} - u^{j-1,k}}{\delta} = \frac{u^{j-1,k+1} - 2u^{j-1,k} + u^{j-1,k-1}}{h^2}, \qquad (6.7)$$

$$k = 1, \ldots, n - 1, \ j = 1, 2, \ldots, N,$$

where

$$\delta = 1/n, \quad and \ h = w/N.$$

(This method is called explicit.)

Problem 90 *Suppose that u, n, f, w, n, N are as in Problem 89 such that (6.4),(6.6) hold. Write a computer program for calculating the quantities in (6.7) using, in place of (6.7), the following scheme (it is called implicit):*

$$\frac{u^{j,k} - u^{j-1,k}}{\delta} = \frac{u^{j,k+1} - 2u^{j,k} + u^{j,k-1}}{h^2}, \tag{6.8}$$

$$k = 1, \ldots, n-1, \; j = 1, 2, \ldots, N.$$

Problem 91 *Observe that it is necessary to solve the system (6.8) for the quantities in (6.5). This is a triangular system. One can use Gaussian elimination, the method of Gauss–Seidel or other methods to solve this system. One can use Mathematica, MatLab, C, Fortran or almost any other computer language. Compare several methods for solving the system in Problem 90.*

Problem 92 *After your codes for Problems 89 and 90 work, make a comparison between the numerical phenomena in Problems 89 and 90 and the analytical phenomena suggested by Problem 84. Think about the first law of numerical analysis in this connection. In particular think about the fact that for $\lambda > 0$, $(I + \lambda A)$ is not continuous but $(I - \lambda A)^{-1}$ is continuous, where A is as in (6.2).*

Problem 93 *Carry out the classical 'separation of variables' method on the heat equation. First determine all solutions $u : [-\pi, \pi] \times [0, \infty)$ so that*

$$u(t, x) = f(t)g(x), \; t \geq 0, x \in [-\pi, \pi],$$
$$\text{with } u(t, -\pi) = 0 = u(t, \pi), \; t \geq 0.$$

Determine that any such nonzero pair must satisfy

$$\frac{f'(t)}{f(t)} = \frac{g''(x)}{g(x)} = \lambda$$

for some $\lambda \in R$, $0 < x < 1$ and $t \geq 0$. Determine all such numbers λ. Show that there is only a countable collection

$$\{\lambda_n\}_{n=0}^{\infty}$$

of such numbers λ. Consider all linear combinations of these 'separated' solutions

$$\sum_{n=1}^{\infty} c_n f_n(t) g_n(x), \; x \in [-\pi, \pi], \; t \geq 0$$

so that

$$\sum_{n=1}^{\infty} c_n^2 \text{ converges. (Why this?)} \tag{6.9}$$

Make a semigroup from this setting.

Problem 94 *Denote by A the transformation with domain all $x = (x_1, x_2, \dots) \in \ell_2$ so that*

$$Ax = (-x_1, -2x_2, -3x_3, -4x_4, \dots) \in \ell_2.$$

Find a semigroup T on ℓ_2 which has generator A.

Problem 95 *Compare semigroups in Problems 93 and 94.*

Chapter 7
Some Problems in Analysis

This chapter has some problems which are preliminary to problems in chapters to follow.

Definition 8 *Suppose each of X and Y is a Banach space and F is a function from a subset Ω of X into Y. The statement that F is Fréchet differentiable at $x \in X$ means that*

- *There is an open set G containing x so that $G \subset \Omega$.*
- *There is $M \in L(X, Y)$ so that if $\epsilon > 0$, there is $\delta > 0$ such that if $h \in H$ and $\|h\|_X < \delta$, then*

$$\|F(x+h) - F(x) - Mh\|_Y \leq \epsilon \|h\|_X, \ \text{if } x + h \in D(F).$$

Problem 96 *Suppose that F is as in Definition 8. Show that the element M in the definition is unique.*

In this case, F' denotes the function whose domain is all $x \in H$ at which F is Fréchet differentiable. For each such $x \in X$, $F'(x)$ denotes the element M in Definition 8.

Definition 9 *A function F' as in Definition 8 is C^1 provided that F' is continuous as a function from $\Omega \to L(X, Y)$.*

Problem 97 *Suppose X is a Banach space $d_0, r > 0$, $(a, b) \in R \times X$ and f is a continuous function from*

$$\Omega = [a - d_0, a + d_0] \times B_r(b) \to X$$

such that for some $M > 0$ it is true that

$$\|f(t, x) - f(t, y)\| \leq M\|x - y\|, \ (t, x), (t, y) \in \Omega.$$

J.W. Neuberger, *A Sequence of Problems on Semigroups*,
Problem Books in Mathematics, DOI 10.1007/978-1-4614-0430-9_7,
© Springer Science+Business Media, LLC 2011

Then there is $d \in (0, d_0)$ and a unique function

$$z : (a - d, a + d) \times X \to X$$

such that

$$z(a) = b, \ z'(t) = f(t, z(t)), \ t \in (a - d, a + d).$$

Consider using the method of successive approximations as in Problem 46.

Problem 98 *Suppose $c > 0$ and h is a function of class C^1 whose domain contains $[0, c)$ and whose range is a subset of the Banach space X. If there is $M > 0$ so that*

$$\int_0^t \|h'\|_X \ \leq \ M, \ t \in [0, c),$$

then

$$\lim_{t \to c-} h(t) \ exists.$$

Problem 99 *Suppose that H is a Hilbert space and f is a continuous linear function from H to R. Show that there is a unique $y \in H$ such that*

$$f(x) = \langle x, y \rangle_H, \ x \in H.$$

Problem 100 *Suppose that H is an infinite-dimensional separable Hilbert space and $T \in L(X, X)$ so that*

$$\langle Tx, y \rangle_H = \langle x, Ty \rangle_H, \ x, h \in H, \tag{7.1}$$

$$\langle Tx, x \rangle_H \geq 0, \ x \in H, \tag{7.2}$$

and if $\{x_k\}^\infty$ is a bounded sequence in H, then the sequence $\{Tx_k\}_{k=0}^\infty$ has a convergent subsequence. Show that there is an orthonormal basis

$$\{\phi_k\}_{k=1}^\infty$$

for H and a nondecreasing sequence in R,

$$\{\lambda_k\}_{k=1}^\infty \tag{7.3}$$

so that

$$T\phi_k = \lambda_k \phi_k, \ k \in Z^+.$$

Problem 101 *Show that the conclusion to Problem 100 still holds if (7.2) is removed and the word 'nondecreasing' is removed where it appears above in (7.3).*

Problem 102 *Suppose that H is a Hilbert space, $T \in L(X,X)$ and (7.1), (7.2) hold. Denote*

$$\sup_{x \in H, \|x\|=1} \langle Tx, x \rangle_H$$

by b.

Show that there is a function ϕ with domain $[0,b]$ and range in the set of orthogonal projections on H so that

- $\phi(0) = 0$.
- *If $0 \leq a < b \leq c < d$ and $x, y \in H$, then*

$$\langle [\phi(b) - \phi(a)]x, [\phi(d) - \phi(c)]x \rangle_H = 0$$

and

$$T = \int_0^b \lambda \, d\phi(\lambda).$$

(The function ϕ is called a spectral family for T - see notes to this chapter.)

Definition 10 *Suppose that T is a closed, densely defined linear transformation on the Hilbert space H into the Hilbert space K. Define T^t as follows: First let*

$$D(T^t) = \{y \in K : \text{ the transformation } W : x \in D(T) \to \langle Tx, y \rangle_K \quad (7.4)$$

is continuous}.

For

$$y \in D(T^t),$$

define

$$T^t y = z$$

where z is the unique element of K such that

$$\langle Tx, y \rangle_K = \langle x, z \rangle_H, \ x \in D(T).$$

Problem 103 *Show that for T as in Definition 10*

$$\langle Tx, y \rangle_K = \langle x, T^t y \rangle, \ x \in D(T), y \in D(T^t),$$

Problem 104 *For T as in Definition 10, show that the range of*

$$(I + T^t T) \text{ is dense in } H.$$

Problem 105 *For T as in Definition 10, show that*

$$\|(I + T^t T)x\|_H \geq \|x\|_H, \ x \in D(T).$$

Problem 106 *Suppose that each of X, Y is a Hilbert space and T is a closed linear transformation on X into Y. Show that*

$$(I + T^tT)^{-1} \in L(X, X)$$

and

$$(I + TT^t)^{-1} \in L(Y, Y).$$

(See [66].)

Problem 107 *For T as in Definition 10, show that*

$$(I + T^tT)^{-1}T^t$$

is continuous and has continuous extension

$$T^t(I + TT^t)^{-1}.$$

Problem 108 *Suppose that each of H, K is a Hilbert space and $T \in L(H, K)$. Denote by T^* the member of $L(K, H)$ so that*

$$\langle Tx, y \rangle_K = \langle x, T^*y \rangle_H, \ x \in H, y \in K.$$

Show that the null space of T is the orthogonal complement of the range space of T^* and that the closure of the range of T is the orthogonal complement of the null space of T^*.

Problem 109 *Suppose that H is a Hilbert space and $T \in L(H, H)$ is self-adjoint, i.e.,*

$$\langle Tx, y \rangle_H = \langle x, Ty \rangle_H, \ x, y \in H$$

and that

$$\langle Tx, x \rangle_H \geq 0, \ x \in H,$$

i.e., T is nonnegative. Show that if $x \in H$, then

$$u = \lim_{t \to \infty} e^{-tT} x \ exists$$

and is the orthogonal projection of x onto the null space of T.

Chapter 8
Combining Semigroups, Linear Continuous Case

Problem 110 *Find two members $A, B \in L(R^2, R^2)$ so that*

$$e^A e^B \neq e^{A+B}.$$

(Use Problem 50 to show that e^A in that problem is equal to

$$\lim_{n \to \infty} (I + \frac{1}{n}A)^n,$$

hence giving two equivalent definitions.)

Problem 111 *Show that if $n \in Z^+, A \in L(R^n, R^n)$ and $\alpha > 0$, then there is $M > 0$ so that*

$$|e^{tA} - (I + tA)| \leq Mt^2, \ t \in [0, \alpha],$$

where

$$|B| = \sup_{x \in R^n, \|x\|=1} \|Bx\|, B \in L(R^n, R^n).$$

Problem 112 *Show that if $n \in Z^+, A \in L(R^n, R^n)$ and $\alpha > 0$, then there is $M > 0$ so that*

$$|e^{tA} - (I + tA + \frac{1}{2!}(tA)^2)| \leq Mt^3, \ t \in [0, \alpha].$$

Problem 113 *Show that if $n \in Z^+$, and $A \in L(R^n, R^n)$, then*

$$|e^A| \leq e^{|A|}.$$

Problem 114 *Suppose that each of A_0, A_1, A_2 and B_0, B_1, B_2 is in $L(R^n, R^n)$. Show that*

$$|A_0 A_1 A_2 - B_0 B_1 B_2| \leq$$
$$|A_0 - B_0||A_1 A_2| + |B_0||A_1 - B_1||A_2| + |B_0 B_1||A_2 - B_2|.$$

J.W. Neuberger, *A Sequence of Problems on Semigroups*,
Problem Books in Mathematics, DOI 10.1007/978-1-4614-0430-9_8,
© Springer Science+Business Media, LLC 2011

Problem 115 *Suppose that each of $m, n \in Z^+$ and each of*

$$\{A_k\}_{k=0}^m \text{ and } \{B_k\}_{k=0}^m \in R^n$$

is in R^n. Generalize the inequality in Problem 114 to this case.

Problem 116 *Suppose that $t > 0$ and $A \in L(R^n, R^n)$. Use results of Problem 115 to estimate*

$$|(I + \frac{t}{m}A)^m - (e^{\frac{t}{m}A})^m|, \ m = 1, 2, \ldots.$$

Problem 117 *Suppose that each of $m, n \in Z^+$ and $t > 0$, $A, B \in L(R^n, R^n)$. Estimate*

$$|((I + \frac{t}{m}A)(I + \frac{t}{m}B))^m - (I + \frac{t}{m}(A + B))^m|.$$

Problem 118 *Suppose that each of $m, n \in Z^+$ and $t > 0$, $A, B \in L(R^n, R^n)$. Estimate*

$$|((I + \frac{t}{m}A)(I + \frac{t}{m}B))^m - (e^{\frac{t}{m}(A+B)})^m|.$$

Problem 119 *Suppose that $n \in Z^+$ and each of $A, B \in L(R^n, R^n)$. Show that*

$$\lim_{m \to \infty} (e^{\frac{t}{m}A}e^{\frac{t}{m}B})^m = e^{A+B}.$$

Definition 11 *In a group G, the commutator between two elements $x, y \in G$ is*

$$(xy)(yx)^{-1} = xyx^{-1}y^{-1}.$$

Definition 12 *In a ring J, the commutator between two elements $A, B \in J$ is*

$$AB - BA.$$

Problem 120 *Suppose that $n \in Z^+, \delta > 0$ and $A, B \in L(R^n, R^n)$. Calculate the following:*

$$(I + \delta A + \frac{\delta^2}{2}A^2)(I + \delta B + \frac{\delta^2}{2}B^2)(I - \delta A + \frac{\delta^2}{2}A^2)(I - \delta B + \frac{\delta^2}{2}B^2),$$

dropping terms of order δ^3 and higher.

Problem 121 *For n, A, B as in Problem 120 and $t > 0$, find an estimate for*

$$|((I + \frac{t}{m}A + \frac{t^2}{2m^2}A^2)(I + \frac{t}{n}B + \frac{t^2}{2m^2}B^2)(I - \frac{t}{m}A + \frac{t^2}{2m^2}A^2)(I - \frac{t}{m}A + \frac{t^2}{2m^2}B^2))^{m^2}$$
$$- (e^{\frac{t}{m}A}e^{\frac{t}{m}B}e^{-\frac{t}{m}A}e^{-\frac{t}{m}B})^{m^2}|.$$

Problem 122 *For n, A, B as in Problem 120 and $t > 0$ show*

$$\lim_{m \to \infty} (e^{\frac{t}{m}A} e^{\frac{t}{m}B} e^{-\frac{t}{m}A} e^{-\frac{t}{m}B})^{m^2} = e^{t^2(AB-BA)}.$$

Problem 123 *Suppose that n, A are as in Problem 122 and $t \geq 0$. Show the following: If $\epsilon > 0$, there is $\delta > 0$ such that if $t_0, t_1, \ldots, t_m$ is a partition from 0 to t, then*

$$|e^{tA} - \Pi_{k=1}^{m}(I + (t_k - t_{k-1})A)| < \epsilon.$$

(Product integral for e^{tA}.)

Problem 124 *Write a product integral formula in the setting of Problem 122.*

Problem 125 *Carry over above results in this chapter with R^n replaced by X, some Banach space.*

Problem 126 *Suppose X is a Banach space and each of $A, B, C \in L(X, X)$. Investigate what conclusions might be drawn about possible*

$$\lim_{n \to \infty} (e^{\frac{t}{n}A} e^{\frac{t}{n}B} e^{\frac{t}{n}C})^n.$$

Problem 127 *Before going on to the next chapter, consider how developments of the present chapter might be generalized to some classes of nonlinear transformations.*

Chapter 9
Combining Semigroups, Nonlinear Continuous Case

Suppose that X is a Banach space and Q is the collection of all transformations $A : X \to X$ such that

- $A0 = 0$.
- A is globally lipschitzian, i.e., there is $M \geq 0$ so that

$$\|Ax - Ay\| \leq M\|x - y\|, \ x, y \in X.$$

- $|A|$ is the least number M in the above item.

If $A, B \in Q$, then AB denotes the composition of the functions A, B.

Problem 128 *Show that the last item above makes a norm for Q.*

Problem 129 *Show that Q as above is a near-normed-ring in the sense that Q has all normed-ring properties except possibly left distributivity and the property that*

$$A(cx) = cAx, \ c \in R, x \in Q.$$

Definition 13 *Define*

$$Q_0 = \{(t, x, A, k) : t \in R, x \in X, A \in Q, k \in Z, k \geq 0\}.$$

Definition 14 *Define*

$$Y : Q_0 \to X$$

so that

$$Y(t, x, A, 0) = x, \ t \in R, x \in X, A \in Q$$

and

$$Y(t, x, A, k) =$$
$$x + \int_0^t AY(\cdot, x, A, k - 1), \ t \in R, x \in X, A \in Q, k \in Z^+.$$

J.W. Neuberger, *A Sequence of Problems on Semigroups*,
Problem Books in Mathematics, DOI 10.1007/978-1-4614-0430-9_9,
© Springer Science+Business Media, LLC 2011

Problem 130 *Show that*

$$\|Y(t, x, A, k+1) - Y(t, x, A, k)\|$$

$$\leq |A| \int_0^t \|Y(\cdot, x, A, k) - Y(\cdot, x, A, k-1)\|,$$

$$t \geq 0, x \in X, A \in Q, k \in Z^+.$$

Problem 131 *For Y as in Definition 14, show inductively that*

$$\|Y(t, x, A, k) - Y(t, x, A, k-1)\|$$

$$\leq \frac{1}{k!}|t|^k|A|^k, \ t \in R, x \in X, A \in Q, k \in Z^+.$$

Problem 132 *For Y as in Definition 14, show that*

$$\{Y(\cdot, x, A, k)\}_{k=0}^\infty$$

converges uniformly on bounded subintervals of R. Define

$$e^{tA} : X \to X$$

so that

$$e^{tA}x = \lim_{k \to \infty} Y(t, x, A, k), \ t \geq 0, x \in X, A \in Q.$$

Problem 133 *Suppose that $A \in Q$ and $t, s \in R$. Show that*

$$e^{tA}e^{sA} = e^{(t+s)A}.$$

Problem 134 *For Y as in Definition 14, show that*

$$|e^{tA}x - Y(t, x, A, k)| \leq \sum_{m=k+1}^\infty \frac{1}{m!}t^{m+1}|A|^{m+1}. \tag{9.1}$$

Problem 135 *Suppose that $A \in Q, x \in X$ and*

$$z(t) = e^{tA}x, \ t \in R.$$

Show that

$$z(0) = x, \ z'(t) = A(z(t)), \ t \in R. \tag{9.2}$$

Problem 136 *Find an instance of X so that the resulting set Q contains an element A so that for some $x \in X$*

$$e^A x \neq \sum_{k=0}^\infty \frac{1}{k!}A^k x.$$

Problem 137 *Suppose that each of*

$$\{A_k\}_{k=1}^n, \{B_k\}_{k=1}^n \in Q.$$

Show that

$$|\Pi_{k=1}^n A_k - \Pi_{k=1}^n B_k|$$
$$\leq |A_1 - B_1||\Pi_{k=2}^n A_k| + |B_1||A_2 - B_2||\Pi_{k=3}^n A_k|$$
$$+ |B_1||B_2||A_3 - B_3||\Pi_{k=4}^n A_k| + \cdots$$
$$+ \Pi_{k=1}^{n-2}|B_k||A_{n-1} - B_{n-1}||A_n| + \Pi_{k=1}^{n-1}|B_k||A_n - B_n|, \ k \geq 1,$$

where all indicated products are taken, from left to right, in the order of increasing subscript.

Problem 138 *Suppose $A \in Q$ and that $t \in R, n \in Z^+$. Let $\delta = \frac{t}{n}$. Apply the inequality in Problem 137 to the case where*

$$A_k = e^{\delta A}, \ B_k = I + \delta A$$

noting that

$$|A_k|, |B_k| \leq e^{|\delta||A|}.$$

Simplify the resulting inequality as much as possible.

Problem 139 *Show that for $A \in Q$,*

$$\lim_{n \to \infty} (I + \frac{t}{n} A)^n = e^t A,$$

the limit being taken in the norm of Q.

Problem 140 *Suppose that $A, B \in Q$ and $\delta > 0$. Show that*

$$|(I + \delta A)(I + \delta B) - (I + \delta(A + B))| \leq \delta^2 |A||B|.$$

Problem 141 *Show that if $A, B \in Q$ and $t \in R$, then*

$$\lim_{n \to \infty} ((e^{\frac{t}{n} A})(e^{\frac{t}{n} B}))^n = e^{t(A+B)}.$$

Problem 142 *Suppose that $A \in Q$ and $t \in R$. Show that if $\epsilon > 0$, there is $\delta > 0$ such that if $t_0, t_1, \ldots, t_n$ is a partition from 0 to t of mesh less than δ, then*

$$|\Pi_{k=1}^n ((e^{(t_k - t_{k-1})A})(e^{(t_k - t_{k-1})B})) - \Pi_{k=1}^n (e^{(t_k - t_{k-1})(A+B)})| < \epsilon.$$

Problem 143 *Suppose $A \in Q$. Show that the semigroup*

$$e^{tA}, \ t \geq 0,$$

extends to a group

$$e^{tA}, \ t \in R,$$

with the property that

$$e^{sA}e^{tA} = e^{(s+t)}A, \quad \text{for all } s, t \in R.$$

Problem 144 *Suppose that each of A, B is a member of Q, that A, B have Fréchet derivatives A', B', respectively, and that there is $M > 0$ with*

$$|A'(x) - A'(y)| \leq M\|x - y\|, \ x, y \in X.$$

Try to show that if $x \in X$ and $t \geq 0$, then

$$\lim_{n \to \infty} (e^{\frac{t}{n}A}e^{\frac{t}{n}B}e^{-\frac{t}{n}A}e^{-\frac{t}{n}B})^{n^2}x = e^{t^2C},$$

where

$$C(x) = A'(x)B(x) - B'(x)A(x), \ x \in X,$$

a nonlinear version of the commutator of A, B given in Problem 122. (An interested reader might see the Notes to this chapter before investing a great deal of time on this problem.)

Problem 145 *Consider that in Problem 134, expression (9.1) is what one gets, using the series expansion for e^{tA} when A is linear. How is it that, in light of Problem 136, the same formula holds in the present nonlinear case?*

Chapter 10
Some Connections Between Resolvents and Linear Semigroups

This group of problems may be thought of as a continuation of those in Chapter 5. A review of Chapter 5 might be in order.

Suppose that X is a Banach space.

Problem 146 *Suppose that each of*

$$Q, \{Q_k\}_{k=1}^{\infty} \in L(X,X) \text{ with } |Q_k| \leq 1, k \in Z^+$$

so that for each $x \in X$,

$$\lim_{k \to \infty} Q_k x = Qx.$$

Show that if $m \in Z^+$, then

$$\lim_{k \to \infty} Q_k^m x = Q^m x, \ x \in X.$$

Problem 147 *Suppose that F is a function whose domain is a dense subset of X so that $G = F^{-1}$ exists, has domain all of X and*

$$\|G(x) - G(y)\| \leq \|x - y\| \text{ for all } x, y \in X.$$

Show that if W is a dense subset of the domain of F, then

$$F(W) \text{ is dense in } X.$$

Problem 148 *Suppose that each of*

$$T, \{T_k\}_{k=1}^{\infty} \tag{10.1}$$

is a linear strongly continuous semigroup of transformations such that

$$|T(t)|, |T_k(t)| \leq 1, \ t \geq 0, k \in Z^+,$$

J.W. Neuberger, *A Sequence of Problems on Semigroups*,
Problem Books in Mathematics, DOI 10.1007/978-1-4614-0430-9_10,
© Springer Science+Business Media, LLC 2011

and that

$$A, \{A_k\}_{k=1}^{\infty}, \tag{10.2}$$

respectively, are the generators of the members of (10.1). *Show that, following the notation of Problem 65,*

$$\|(I - \lambda A)^{-k}x - (I - \lambda A_n)^{-k}x\| \leq \int_0^{\infty} d\phi_{k,\lambda} \, \|T(\cdot)x - T_n(\cdot)x\|,$$

for $n, k \in Z^+$, $x \in X$.

Problem 149 *In addition to the hypothesis of Problem 148, suppose that if* $x \in X$, *then*

$$T_n(\cdot)x$$

converges to

$$T(\cdot)x \text{ as } n \to \infty,$$

uniformly on each bounded subinterval of $[0, \infty)$. *Show that*

$$\lim_{n \to \infty} (I - \lambda A_n)^{-k}x = (I - \lambda A)^{-k}x, \ k \in Z^+, x \in X.$$

Problem 150 *In addition to the hypothesis of Problem 148, suppose that* W *is a dense subset of* X *which is a subset of the domain of each of the transformations in* (10.2). *Suppose finally that for each* $x \in W$,

$$\lim_{k \to \infty} A_k x = Ax.$$

Show that for each $\lambda > 0$ *and* y *in the range of* $(I - \lambda A)$,

$$\lim_{k \to \infty} (I - \lambda A_k)^{-1}y = (I - \lambda A)^{-1}y.$$

Problem 151 *Under the hypothesis of Problem 150, show that if* $m \in Z^+$, $x \in X$ *and* $\lambda > 0$, *then*

$$\lim_{k \to \infty} (I - \lambda A_k)^{-m}x = (I - \lambda A)^{-m}x.$$

Problem 152 *Under the hypothesis of Problem 150, show that if* $x \in W$, *then*

$$\{A_n x\}_{n=1}^{\infty}$$

is a bounded sequence.

Problem 153 *Under the hypothesis of Problem 148, show that if* $x \in W$, *then*

$$(T_n(\cdot)x)'(t) = AT_n(t)x = T_n(t)Ax, \ t \geq 0, \ n \in Z^+.$$

Problem 154 *Under the hypothesis of Problem 150, show that if $x \in W$, then*

$$\{T_n(\cdot)x\}_{n=1}^\infty$$

is a bounded and equicontinuous collection on each bounded subinterval of $[0, \infty)$.

Problem 155 *Show that if $\lambda > 0$, there is $M \in Z^+$ so that if $m > M$, then there are maximal $a_m < \lambda$, minimal $b_m > \lambda$ so that*

$$0 < \frac{1}{2} - \int_{a_m}^{\lambda} d\phi_{m,\lambda} \le \epsilon \text{ and } 0 < \frac{1}{2} - \int_{\lambda}^{b_m} d\phi_{m,\lambda} \le \epsilon.$$

(See Problem 65 for notation.)

Problem 156 *Show that for Problem 155,*

$$b_m - a_m \to 0 \text{ as } m \to \infty.$$

Problem 157 *Suppose $x \in W$. Assuming the denial of the conclusion to Problem 160 show that there is*

- $\epsilon > 0$, a bounded subinterval $[a, b]$ of $[0, \infty)$
- *a member λ of (a, b)*
- *a sequence $\{\lambda_j\}_{j=1}^\infty$ converging to λ*
- *an increasing sequence $\{n_j\}$ of positive integers*

so that
$$\|T_j(\lambda_{n_j})x - T(\lambda_{n_j})x\| > \epsilon, \; j \in Z^+.$$

Problem 158 *Using the assumptions, conclusions of Problem 157 and the result of Problem 154, show that there are*

- *a sequence $\{a_j\}_{j=1}^\infty$ converging monotonically increasing to λ*
- *a sequence $\{b_j\}_{j=1}^\infty$ converging monotonically decreasing to λ*

so that
$$\left\| \int_{a_j}^{b_j} (T_{n_j}(\cdot)x - T(\cdot)x) \right\| \ge \epsilon, \; j \in Z^+.$$

Problem 159 *Continuing on with Problem 158 and using the results of Problem 154, reach a contradiction.*

Problem 160 *Under the hypothesis of Problem 150, show that for each $x \in W$,*

$$T_n(\cdot)x$$

converges uniformly on every bounded subinterval of $[0, \infty)$ to

$$T(\cdot)x \text{ as } n \to \infty.$$

Problem 161 *Under the hypothesis of Problem 150, show that for each* $x \in X$,

$$\{T_n(\cdot)x\}_{n=1}^{\infty}$$

converges uniformly on every bounded subinterval of $[0, \infty)$ *to*

$$T(\cdot)x \text{ as } n \to \infty.$$

Problem 162 *Study [17] Chapter III, Section 5 to see an alternate path to Problem 161.*

Chapter 11
Combining Semigroups, Strongly Continuous Linear Case

Suppose that X is a Banach space.

Problem 163 *Suppose that each of*

$$\{T_k\}_{k=1}^\infty \tag{11.1}$$

is a linear strongly continuous semigroup of transformations such that

$$|T_k(t)| \leq 1,\ t \geq 0, k \in Z^+.$$

Suppose also that

$$\{A_k\}_{k=1}^\infty,$$

respectively, are the generators of the members of (11.1). Suppose also that A is a densely defined linear transformation on X such that

$$\lim_{k \to \infty} A_k x = Ax,\ x \text{ in some dense subset of the range of } A.$$

Suppose finally that

$$\text{the range of } I - \lambda A \text{ is dense in } X$$

for all $\lambda > 0$. Show that if $\lambda > 0$, then

$$(I - \lambda A_k)^{-1}x \text{ converges, as } k \to \infty, \tag{11.2}$$

for all x in the range of $I - \lambda A$.

Problem 164 *Under the hypothesis of Problem 163 show that*

$$(I - \lambda A_k)^{-1}x \text{ converges for all } x \in X \text{ as } k \to \infty. \tag{11.3}$$

Problem 165 *Under the hypothesis of Problem 163 show that if $m \in Z^+$, then*

$$(I - \lambda A_k)^{-m}x \text{ converges for all } x \in X \text{ as } k \to \infty. \tag{11.4}$$

J.W. Neuberger, *A Sequence of Problems on Semigroups*,
Problem Books in Mathematics, DOI 10.1007/978-1-4614-0430-9_11,
© Springer Science+Business Media, LLC 2011

Problem 166 *Under the hypothesis of Problem 163 show that if $j, k, m \in Z^+$, then*

$$(I - \frac{\lambda}{m}A_j)^{-m}x - (I - \frac{\lambda}{m}A_k)^{-m}x = \frac{1}{\lambda}\int_0^\infty d\phi_{m,\lambda}\,(T_j(\cdot)x - T_k(\cdot)x)$$

for all $x \in X$, $\lambda > 0$.

Problem 167 *Show that under the hypothesis of Problem 166, if $x \in X$, then*

$$T_1(\cdot)x, T_2(\cdot)x, \ldots \tag{11.5}$$

converges uniformly on each bounded subinterval of $[0, \infty)$. (An argument by way of contradiction might be considered here.)

Under the hypothesis of Problem 167, denote by T the function with domain $[0, \infty)$ and range the collection of functions from X to X so that if $x \in X$ and $\lambda \geq 0$, then

$$T(\lambda)x = \lim_{k \to \infty} T_k(\lambda)x.$$

Problem 168 *Show that*

$$T(\lambda) \in L(X, X), \; \lambda \geq 0.$$

Problem 169 *Show that if $x \in X$, then*

$$T(\cdot)x$$

is continuous.

Problem 170 *Show that*

$$|T(\lambda)| \leq 1, \; \lambda \geq 0.$$

Problem 171 *Show that*

$$T(0) = I, \; T(t)T(s) = T(t + s), \; t, s \geq 0.$$

Problem 172 *Conclude that T is a linear, nonexpansive, strongly continuous semigroup on X.*

Problem 173 *Suppose that each of K and S is a linear, nonexpansive, strongly continuous semigroup on X with generators B, C, respectively. Suppose that*

- $W = Domain(B) \cap Domain(C)$ *is dense in X.*
- $I - \lambda(B + C)$ *has range dense in X if $\lambda \geq 0$.*
-

$$A_n = n((K(\frac{1}{n})S(\frac{1}{n}) - I), \; n \in Z^+.$$

Find A so that

$$Ax = \lim_{n \to \infty} A_n x, \ x \in W.$$

Problem 174 *Under the conditions of Problem 173 show that the sequence*

$$A_1, A_2, \ldots$$

gives rise to a semigroup T as in Problem 167.

Problem 175 *Investigate relationships between the generators B, C of Problem 173 and the generator A of T in Problem 174.*

Problem 176 *Note that the above problems in this chapter involve nonexpansive semigroups. Develop a corresponding theory for general strongly continuous linear semigroups T for which there is $M > 0$ so that*

$$|T(t)| \le M, \ t \ge 0.$$

Problem 177 *Note that the above problems in this chapter involve nonexpansive semigroups. Develop a corresponding theory for general strongly continuous linear semigroups T for which there are $M, \omega > 0$ so that*

$$|T(t)| \le M e^{wt}, \ t \ge 0.$$

Problem 178 *Read Chapter III, Section 5 of [17] for stronger results than given above in this chapter. Also read the discussion in [19] concerning Trotter–Kato formulae and related developments concerning the Feynman–Kac formula. Examine the development in this chapter and make comparisons with [17],[19]. See Notes for this chapter for additional comments in this regard.*

Problem 179 *Suppose that*

$$\{T_n\}_{k=n}^{\infty}$$

is a sequence of nonexpansive linear strongly continuous semigroups on the Banach space X for which there is a dense subset W of X common to its corresponding sequence of generators

$$\{A_n\}_{n=1}^{\infty}.$$

Suppose also that if $x \in W$, then

$$\lim_{n \to \infty} A_n x$$

exists. Try to determine if there is a strongly continuous linear nonexpansive semigroup T such that if $x \in X$, then

$$\{T_n(\cdot)x\}_{n=1}^{\infty}$$

converges uniformly on each closed and bounded subinterval of $[0, \infty)$.

Problem 180 *Suppose that each of T and S is a strongly continuous linear nonexpansive semigroup on the Banach space X. Try to determine if there is a strongly continuous linear nonexpansive semigroup U on X so that*

$$U(t)x = \lim_{n \to \infty} (T(\frac{t}{n})S(\frac{t}{n}))^n x, \ x \in X.$$

Problem 181 *In addition to the conditions of Problem 180 suppose that A, B are the generators of T, S respectively, $\lambda > 0$ and*

$$A_n = \frac{n}{\lambda}((I - \frac{\lambda}{n}A)^{-1} - I), B_n = \frac{n}{\lambda}((I - \frac{\lambda}{n}B)^{-1} - I),$$

for $n \in Z^+$. Denote by

$$\{T_n\}_{n=1}^{\infty}, \ \{S_n\}_{n=1}^{\infty}$$

two sequences of continuous linear nonexpansive semigroups generated

$$\{A_n\}_{n=1}^{\infty}, \ \{B_n\}_{n=1}^{\infty},$$

respectively. Finally define

$$\{U_n\}_{n=1}^{\infty}$$

by

$$U_n(t)x = \lim_{k \to \infty} ((T_n(\frac{t}{k})(S_n(\frac{t}{k}))^k x, \ x \in X, n \in Z^+.$$

Is it true that

$$\{U_n\}_{n=1}^{\infty}$$

is such that there is a strongly continuous linear semigroup U on X such that if $x \in X$, then

$$\{U_n(\cdot)x\}_{n=1}^{\infty}$$

converges uniformly to

$$U(\cdot)x$$

on each closed and bounded subset of $[0, \infty)$?

Problem 182 *Find and read [23] for generalization of Trotter-Kato results to nonlinear semigroups.*

Chapter 12
Splitting Method, Numerics

Developments in the preceding chapters suggest a practical technique for time-dependent partial differential equations. Suppose $w > 0$ and $X = L_2([0,1])$ and each of A, B, C is a linear transformation from a dense subset of X into X. For some $w > 0$ one might seek

$$u : [0, w] \times X \to X$$

so that

$$u_1(t, x) = A(u(t, \cdot))(x) + B(u(t, \cdot))(x) + C(u(t, \cdot))(x), \qquad (12.1)$$

$t \in [0, w]$, $x \in [0, 1]$, where u_1 denotes the partial derivative of u in its first argument.

Problem 183 *Show how, with $B = C = 0$, A may be chosen so that (12.1) is the heat equation of Chapter 6.*

Problem 184 *Show how to choose A, B, C in (12.1) so that the equation becomes*

$$u_1(t, x) = u_{2,2}(t, x) + u_2(t, x) + u(t, x), \; t \in [0, w], x \in [0, 1]. \qquad (12.2)$$

Problem 185 *Devise a numerical scheme for solving*

$$u_1(t, x) = u_{2,2}(t, x), \; t \in [0, w], \; x \in [0, 1].$$

Problem 186 *Devise a numerical scheme for solving*

$$u_1(t, x) = u_2(t, x), \; t \in [0, w], \; x \in [0, 1].$$

Problem 187 *Devise a numerical scheme for solving*

$$u_1(t, x) = u(t, x), \; t \in [0, w], \; x \in [0, 1].$$

J.W. Neuberger, *A Sequence of Problems on Semigroups*,
Problem Books in Mathematics, DOI 10.1007/978-1-4614-0430-9_12,
© Springer Science+Business Media, LLC 2011

Problem 188 *Suppose one seeks u satisfying* (12.2) *with the boundary condition*

$$u(t, 0) = 0, \ t \in [0, w].$$

Define semigroups T_1, T_2, T_3 *from Problems 185, 186, 187, respectively. Determine if the extension of Problem 173 to the combination of three semigroups might allow a formula for solving* (12.2) *by means of a finite-dimensional version of*

$$(T_1(\frac{t}{n})T_2(\frac{t}{n})T_3(\frac{t}{n}))^n f, \ t \in [0, w] \tag{12.3}$$

where

$$u(0, x) = f(x), \ x \in [0, 1],$$

f being a given member of $L_2([0, 1])$ *specifying an initial condition for* (12.2).

Problem 189 *Code your solutions to Problems 185, 186, 187. Consider using an implicit scheme for the part coming from Problem 185.*

Problem 190 *Use the main parts of your codes in Problem 189 to implement* (12.3). *Test for various choices of n and various discretizations of* $[0, 1]$.

Equation (12.2) is a linear example of a reaction–diffusion–convection equation. The term corresponding to A is the diffusion term, the one corresponding to B is the convection term and the one corresponding to C is the reaction term.

This splitting method is often used in cases where there is not yet a proof of convergence, but is often used to great success in a practical way. A typical example is indicated by the following:

Problem 191 *Suppose that* $w > 0$. *Devise a code to implement a splitting method to find a numerical approximation to* u, v *such that*

$$u_1(t, x, y) = b_1 u_{2,2}(t, x, y) + b_2 u_{3,3}(t, x, y)$$
$$+ c_1 u_2(t, x, y) + d_1 u_3(t, x, y) + m_1 u(t, x, y)v(t, x, y),$$
$$v_1(t, x, y) = b_2 v_{2,2}(t, x, y) + v_{3.3}(t, x, y)$$
$$+ c_2 v_2(t, x, y) + d_2 v_3(t, x, y) + m_2 u(t, x, y)v(t, x, y),$$
$$t \in [0, w], \ x, y \in [0, 1] \times [0, 1],$$

where $b_1, b_2, c_1, c_2 > 0, \quad d_1, d_2, m_1, m_2 \in R$ *and*

$$u(t, 0, 0) = u(t, 1, 0) = u(t, 0, 1) = u(t, 1, 1), \ t \geq 0,$$

$$u(0, x, y) = h(x, y), \ (x, y) \in [0, 1]^2$$

for some given function h on $[0, 1]$ *specifying initial conditions.*

Problem 192 *Define A, B, C so that with*

$$z = \begin{pmatrix} u \\ v \end{pmatrix}$$

the equation in Problem 191 becomes

$$z' = Az + Bz + Cz.$$

Chapter 13
Semigroups of Steepest Descent, Abstract Linear Case

Problem 193 *Suppose that $m, n \in Z^+$ and $A \in L(R^n, R^m)$. Suppose also that*

$$\binom{r}{s} \in R^n \times R^m$$

and

$$\phi(x) = \frac{1}{2}\|\binom{x}{Ax} - \binom{r}{s}\|^2_{(R^n, R^m)}.$$

Find the minimum of ϕ.

Problem 194 *Suppose that H, K are two Hilbert spaces, $G \in L(H, K)$ and $h \in K$. Define*

$$\phi(x) = \frac{1}{2}\|Gx - h\|^2_K, \ x \in H.$$

Denote by G^* the element of $L(K, H)$ such that

$$\langle Gx, y\rangle_K = \langle x, G^*y\rangle_H, \ x \in H, \ y \in K.$$

Show that

$$(\nabla\phi)(x) = G^*Gx - G^*h, \ x \in H$$

where $(\nabla\phi)(x)$ is the element of H so that

$$\phi'(x)h = \langle h, (\nabla\phi)(x)\rangle_H, \ x, h \in H.$$

Problem 195 *For $\nabla\phi$ as in Problem 194, $x \in H$ and z the unique solution of*

$$z(0) = x, \ z'(t) = -(\nabla\phi)(z(t)), \ t \geq 0,$$

show that

$$z(t) = e^{-tG^*G}x + \int_0^t e^{-(t-s)G^*G}G^*h \, ds. \tag{13.1}$$

(Variation of parameters for ordinary differential equations.)

J.W. Neuberger, *A Sequence of Problems on Semigroups*,
Problem Books in Mathematics, DOI 10.1007/978-1-4614-0430-9_13,
© Springer Science+Business Media, LLC 2011

Denote by Q the orthogonal projection of H onto the null space of G and denote by P the orthogonal projection of K onto the range of G.

Problem 196 *In the setting of Problem 195 show that*

$$\lim_{t\to\infty} e^{-tG^*G}x = Qx.$$

Problem 197 *In the setting of Problem 195 show that if $t \in R$, then*

$$Ge^{tG^*G} = e^{tGG^*}G \text{ and } G^*e^{GG^*} = e^{G^*G}G^*.$$

Problem 198 *For z as in Problem 195 show that*

$$\lim_{t\to\infty} G(z(t)) = g$$

where g is the orthogonal projection of h onto the range of G.

Problem 199 *For z as in Problem 195 show that if $h \in range\ G$, then*

$$u = \lim_{t\to\infty} z(t) \text{ exists and } Gu = h.$$

Problem 200 *Suppose that z is as in Problem 199, and $y \in H$ is such that $Gy = h$. Show that*

$$\|u - z(t)\| \leq \|y - z(t)\|,\ t \geq 0.$$

Problem 201 *Consider an alternative to using continuous steepest descent for finding a zero of G:*

Suppose that ϕ, h, G are as in Problem 194 and h is in the range of G. Suppose also that $u_0 \in X$, and

$$u_k = u_{k-1} - \delta_k(\nabla\phi)(u_{k-1}),\ k = 1, 2, \ldots$$

where at step $k \in Z^+$,

$$\nabla\phi_k \text{ is from } (13.1)$$

and δ_k is chosen to minimize q:

$$q(t) = \min \|\phi(u_{k-1} - t(\nabla\phi)(u_{k-1}))\|_X,\ t \geq 0.$$

Show that the sequence $\delta_1, \delta_2, \ldots$ is unique. Show that $u_0, u_1, \ldots$ converges to u such that

$$\phi(u) \text{ is minimum.}$$

This illustrates discrete steepest descent for linear problems.

The following is both an application to the developments of this chapter and a preview of Chapters 14 and 15.

Problem 202 *Suppose that*

$$\Omega = [-1,1]^2$$

and X, Y are the Sobolev spaces

$$H^{2,2}(\Omega), L_2(\Omega),$$

respectively. Denote by $G : X \to Y$ the transformation so that

$$(Gu)(x,y) = yu_{1,1}(x,y) + u_{2,2}(x,y), \ (x,y) \in \Omega.$$

Show that the problem of finding

$$Gu = 0 \qquad\qquad (13.2)$$

is elliptic, hyperbolic in

$$[-1,1] \times (0,1] \ and \ [-1,1] \times [-1,0),$$

respectively. The problem of solving (13.2) is called a Tricomi problem.

Do you know what boundary or supplementary conditions to impose are necessary and sufficient in order that there is one and only one solution to this problem on Ω? It is my understanding that this problem has not been solved.

Problem 203 *After considering Chapter 15 write a code for an appropriate discretization of Problem 202. Observe how an answer to an iteration as in Problem 202 depends on the choice of u_0.*

Chapter 14
Semigroups of Steepest Descent for Differential Equations

The first problem in this chapter seeks to make the point that for a given linear transformation A on a finite-dimensional space to itself, an adjoint for A depends on a choice of inner products, one for the domain space and one for the range space.

Problem 204 *Suppose that $A \in L(R^2, R^2)$ defined by*

$$A\begin{pmatrix} x \\ y \end{pmatrix} = \begin{pmatrix} x + 2y \\ 3x + 4y \end{pmatrix}, \quad \begin{pmatrix} x \\ y \end{pmatrix} \in R^2.$$

Suppose also that in addition to the standard inner product $\langle \cdot, \cdot \rangle_{R^2}$, one has a second inner product $\langle \cdot, \cdot \rangle_S$ defined by

$$\left\langle \begin{pmatrix} r \\ s \end{pmatrix}, \begin{pmatrix} u \\ v \end{pmatrix} \right\rangle_S = \left\langle \begin{pmatrix} r \\ s \end{pmatrix}, \begin{pmatrix} u \\ v \end{pmatrix} \right\rangle_{R^2} + (r - s)(u - v), \quad \begin{pmatrix} u \\ v \end{pmatrix}, \begin{pmatrix} r \\ s \end{pmatrix} \in R^2.$$

Find a linear transformation $B \in L(R^2, R^2)$ such that

$$\left\langle A\begin{pmatrix} u \\ v \end{pmatrix}, \begin{pmatrix} r \\ s \end{pmatrix} \right\rangle_{R^2} = \left\langle \begin{pmatrix} u \\ v \end{pmatrix}, B\begin{pmatrix} r \\ s \end{pmatrix} \right\rangle_S, \quad \begin{pmatrix} u \\ v \end{pmatrix}, \begin{pmatrix} r \\ s \end{pmatrix} \in R^2.$$

For the remainder of this chapter H denotes a Hilbert space. There are two objectives for the problems in this chapter. One is to describe an important class of semigroups. The other is to introduce a theory of steepest descent for partial differential equations.

Problem 205 *Show that if f is a continuous linear function from H to R (that is to say, a member of the dual space H^* of H), then there is a unique $y \in H$ so that*

$$f(x) = \langle x, y \rangle_H, \quad x \in H. \tag{14.1}$$

J.W. Neuberger, *A Sequence of Problems on Semigroups*,
Problem Books in Mathematics, DOI 10.1007/978-1-4614-0430-9_14,
© Springer Science+Business Media, LLC 2011

Definition 15 *Suppose that ϕ is a C^1 function from $H \to R$. The gradient of ϕ is the function $\nabla\phi : H \to R$ so that*

$$\phi'(x)h \,=\, \langle h, (\nabla\phi)(x)\rangle_H, \ x, h \in \ H.$$

For the rest of this chapter we suppose that the gradient $\nabla\phi$ is defined on all of H and $\nabla\phi$ is locally lipschitz, that is, if $x \in H$ there is $\delta, M > 0$ such that

$$\|(\nabla\phi)(w) - (\nabla\phi)(y)\|_H \,\leq\, M\|w - y\|_H$$

if $\|w - x\|, \|y - x\| \leq \delta$.

Problem 206 *Suppose that $w > 0$, $x \in H$ and $z : [0, w] \to H$ so that*

$$z(0) = x, \ z'(t) = -(\nabla\phi)(z(t)), \ t \in [0, w].$$

Show that

$$(\phi(z))'(t) \,=\, -\|(\nabla\phi)(z(t))\|^2, \ t \in [0, w].$$

Problem 207 *Show that, given $x \in H$, there is a unique function z : $[0, \infty) \to H$ such that*

$$z(0) = x, \ z'(t) \,=\, -(\nabla\phi)(z(t)), \ t \in [0, \infty). \tag{14.2}$$

Denote by T_ϕ the semigroup generated by (14.2), i.e., if $x \in H$ and $s \geq 0$, then

$$T_\phi(s)x = z(s),$$

where z satisfies (14.2).

Problem 208 *Show that if $x \in H$ and*

$$u \,=\, \lim_{t\to\infty} T_\phi(t)x \ exists, \tag{14.3}$$

then

$$(\nabla\phi)(u) = 0.$$

The study of limits in (14.3) is very important in the theory of semigroups. For many problems in the theory of differential equations in variational form (represented by a function ϕ) the critical points of ϕ are the solutions to the problem. In the notes in the last chapter there are additional references to applications. If a system of equations does not arise from a conventional variational form, one may often construct a function ϕ such that its zeros are solutions. This is illustrated by means of the following sequence of problems devoted to one of the simplest possible examples cast into a variational form: Find u with domain $[0, 1]$ so that

$$u' - u = 0. \tag{14.4}$$

We know that u satisfies (14.4) if and only if there is $c \in R$ such that

$$u(t) = ce^t, \ t \in [0, 1]$$

but it is good to study a new method in a simple known case.

We can try to place (14.4) in a variational form by introducing ϕ such that

$$\phi(u) = \frac{1}{2} \int_0^1 (u' - u)^2, \ u \in H. \tag{14.5}$$

But how do we choose H?

Problem 209 *Show that if $H = L_2([0, 1])$, then ϕ has as its domain a linear set only dense in H. In addition, show that ϕ is nowhere continuous.*

Problem 210 *What do you think about the choice $H = L_2([0, 1])$ for a space on which to try to minimize Φ? Would a useful gradient be forthcoming if this choice is made?*

The next problems introduce a Sobolev space which will serve us well. It is the simplest example of a Sobolev space, but the ideas in these problems carry over to very general cases. This provides an alternate, but equivalent, definition for $H^{1,2}([0, 1])$ given in Chapter 7.

Denote by G_1 the set

$$\left\{ \begin{pmatrix} u \\ u' \end{pmatrix} : u \in C^1([0, 1]) \right\}.$$

Problem 211 *Show that G_1 is a linear subspace of $L_2([0, 1])^2$ with norm*

$$\left\| \begin{pmatrix} f \\ g \end{pmatrix} \right\|_{L_2([0,1])^2} = (\|f\|^2 + \|g\|^2)^{1/2}, \ f, g \in C^1([0, 1]). \tag{14.6}$$

Problem 212 *Denote by G_2 the closure, in $L_2([0, 1])^2$, of G_1. Show that there are not two members of G_2 with the same first term.*

Definition 16

$$H = H^{1,2}([0, 1])$$

denotes the space of all first terms of members of G_2. If $f \in H$ with $\begin{pmatrix} f \\ g \end{pmatrix} \in G_2$, write f' for g and say that g is the generalized derivative of f. For norm in this space H take

$$\|f\|_H = (\|f\|^2_{L_2([0,1])} + \|f'\|^2_{L_2([0,1])})^{1/2}.$$

Problem 213 *Can one justify defining g in the above definition as f'? Show that if $f \in C^1([0, 1])$, then this definition is consistent with the usual one.*

Problem 214 *Show that if $f \in H^{1,2}$, then for some number c,*

$$f(x) = c + \int_0^x f', \ x \in [0,1].$$

Problem 215 *Show that ϕ in (14.5) is continuous on H.*

Problem 216 *Find ϕ' for ϕ as in (14.5).*

We want to find an expression for $\nabla\phi$ where ϕ is defined in (14.5).

Problem 217 *Show that if*

$$\begin{pmatrix} w \\ v \end{pmatrix} \in (L_2([0,1]))^2$$

and

$$\left\langle \begin{pmatrix} u \\ u' \end{pmatrix}, \begin{pmatrix} w \\ v \end{pmatrix} \right\rangle_{L_2([0,1])^2} = 0, \ \begin{pmatrix} u \\ u' \end{pmatrix} \in G_2,$$

then

$$v \in H, \ w = v' \text{ and } v(0) = 0 = v(1).$$

Problem 218 *Construct the projection P of all of $L_2([0,1])^2$ onto*

$$\left\{ \begin{pmatrix} u \\ u' \end{pmatrix} : u \in H \right\}.$$

To do this, take $\begin{pmatrix} f \\ g \end{pmatrix} \in L_2([0,1])^2$. Seek $u \in H$ such that

$$\left\| \begin{pmatrix} u \\ u' \end{pmatrix} - \begin{pmatrix} f \\ g \end{pmatrix} \right\|^2_{L_2([0,1])}$$

is minimum, that is, seek $u, v \in H$ such that

$$\begin{pmatrix} u \\ u' \end{pmatrix} + \begin{pmatrix} v' \\ v \end{pmatrix} = \begin{pmatrix} f \\ g \end{pmatrix} \text{ and } v(0) = 0 = v(1). \tag{14.7}$$

In the notes there are references which contain extensive information on projections which one encounters in the construction of Sobolev gradients.

Problem 219 *Solve the system (14.7). Note that, thanks to Problem 217, there is only one pair (u, v) which satisfies (14.7). Define*

$$S(t) = \sinh(t); \ C(t) = \cosh(t), \ t \in R,$$

show that

$$u(t) = [C(1-t) \int_0^t (C(r)f(r) + S(r)g(r))\, dr +$$

$$C(t) \int_t^1 (C(1-r)f(r) - S(1-r)g(r))\, dr]/S(1), \ t \in [0,1].$$

Problem 220 *With ϕ as in (14.5), P as in Problem 218 and*

$$\pi : L_2([0,1])^2 \to L_2([0,1])$$

defined by

$$\pi\begin{pmatrix} f \\ g \end{pmatrix} = f, \ \begin{pmatrix} f \\ g \end{pmatrix} \in L_2([0,1])^2,$$

show that

$$(\nabla\phi)(y) = \pi P\begin{pmatrix} y - y' \\ y' - y \end{pmatrix}, \ y \in H. \tag{14.8}$$

Problem 221 *Find a simple form for $\nabla\phi$ in Problem 220.*

Problem 222 *Search for an expression for the solution z of (14.2) using the gradient in Problem 221 and search for a form for u in (14.3).*

Problem 223 *Show that in Problem 221 a limit depends on the selection of $z(0)$ in (14.2).*

Problem 224 *Show that a limit u in Problem 221 is the nearest element to $z(0)$ in the norm of $H^{1,2}([0,1])$ where z is as in (14.2).*

Usually one cannot find an explicit form for the Sobolev gradient for ϕ. For many cases one can use the above ideas to try to prove that the limit u in (14.3) exists and is a zero of ϕ. In addition, one may try to follow a trajectory z numerically. Some of the problems which follow deal with existence of the limit u in (14.3).

Definition 17 *Suppose ϕ is a function from a Hilbert space H into $[0, \infty)$ of class C^1 with a locally lipschitzian gradient and $\Omega \subset H$. One says that ϕ satisfies a gradient inequality on Ω if there is $c > 0$ such that*

$$\|(\nabla\phi)(x)\|_H \geq c(\phi(x))^{1/2}, \ \ if \ x \in \Omega. \tag{14.9}$$

Problem 225 *Suppose H is a Hilbert space, ϕ a function from H to $[0, \infty)$ such that $\nabla\phi$ is locally lipschitz, $x \in H$, and z the unique solution of*

$$z(0) = x, \ z'(t) = -(\nabla\phi)(z(t)), \ t \geq 0.$$

Suppose also that $\Omega \subset H$ is such that ϕ satisfies a gradient inequality (14.9) in Ω with constant c. Show that if

$$range(z) \subset \Omega,$$

then

$$(\phi(z))'(t) \leq -c^2\phi(z(t)), \ t \geq 0. \tag{14.10}$$

Problem 226 *Show that if (14.10) holds, then*

$$\phi(z(t)) \leq \phi(z(a))e^{-c^2(t-a)}, \ t \geq a.$$

Problem 227 *Show that for z as in Problem 225,*

$$u = \lim_{t \to \infty} z(t) \ exists$$

and

$$\phi(u) = 0.$$

Chapter 15
Numerics for Semigroups of Steepest Descent

At first we work with some numerical problems. We use the same example as in Chapter 14 but in a discrete form.

Suppose $n > 2$ is an integer. Define

$$\delta = 1/n.$$

Suppose

$$\phi_n : R^{n+1} \to R$$

such that

$$\phi_n(u_0, u_1, \ldots, u_n) = \frac{1}{2} \sum_{k=0}^{n} \left(\frac{u_k - u_{k-1}}{\delta} - \frac{u_k + u_{k-1}}{2} \right)^2,$$

with $(u_0, u_1, \ldots, u_n) \in R^{n+1}$.

Problem 228 *Find a formula for*

$$\nabla \phi_n,$$

the conventional gradient of ϕ_n, that is, a vector of length $n + 1$ of partial derivatives of ϕ_n.

Problem 229 *Show that*

$$\phi_n'(u)h = \langle h, (\nabla \phi_n)(u) \rangle_{R^{n+1}}, \ h, u \in R^{n+1}. \tag{15.1}$$

Problem 230 *If*

$$u = (u_0, u_1, \ldots, u_n) \in R^{n+1},$$

find the unique number $\alpha_{n,u}$ such that

$$\phi_n(u - \alpha_{n,u}(\nabla \phi_n)(u)) \ is \ minimum.$$

(Find an explicit expression using the usual inner product in R^{n+1}.)

J.W. Neuberger, *A Sequence of Problems on Semigroups*,
Problem Books in Mathematics, DOI 10.1007/978-1-4614-0430-9_15,
© Springer Science+Business Media, LLC 2011

Problem 231 *Show that the iteration*

$$u \to u - \alpha_{n,y}(\nabla\phi_n)(u) \tag{15.2}$$

converges to the limit u and that

$$\phi_n(u) = 0.$$

Problem 232 *Write a computer program which follows the iteration in Problem 231. (Choose $n = 10, n = 20, n = 100$ and print or graph your results.)*

Problem 233 *Run your code developed in Problem 232. Notice that your code requires many iterations if the integer n is even as much as 20. Reflect on the First Law of Numerical Analysis, just before Problem 89. Reflect also Problems 83, 84. What is going on?*

Definition 18 *Define a second norm on R^{n+1}, called $\| \cdot \|_{S_n}$:*

$$\|u\|_{S_n} = \left(\sum_{k=1}^{n}\left(\frac{u_k - u_{k-1}}{\delta}\right)^2 + \left(\frac{u_k + u_{k-1}}{2}\right)^2\right)^{1/2},$$

imitating the norm in $H^{1,2}([0,1])$ in (14.6).

Definition 19 *Define two linear transformations D_0, D_1*

$$D_0, D_1 : R^{n+1} \to R^n$$

such that if

$$u = (u_0, u_1, \ldots, u_n) \in R^{n+1},$$

then

$$D_0 u = \left\{\frac{u_1 + u_0}{2}, \ldots, \frac{u_n + u_{n-1}}{2}\right\}$$

and

$$D_1 u = \left\{\frac{u_1 - u_0}{\delta}, \ldots, \frac{u_n - u_{n-1}}{\delta}\right\}.$$

Definition 20

$$\nabla_{S_n}\phi_n$$

is the function $R^{n+1} \to R^{n+1}$ so that

$$\phi_n'(u)h = \langle h, (\nabla_{S_n}\phi_n)(u)\rangle_{S_n}, \ h, u \in R^{n+1}.$$

(One can represent the linear function $R^{n+1} \to R$ in any inner product defined on R^{n+1}.)

Definition 21 *Suppose D is the transformation*

$$R^{n+1} \to (R^n)^2$$

such that

$$Du = \begin{pmatrix} D_0 u \\ D_1 u \end{pmatrix}, \ u \in R^{n+1}.$$

Problem 234 *Show that if $u, v \in R^{n+1}$, then*

$$\langle u, v \rangle_{S_n} = \langle Du, Dv \rangle_{(R^n)^2}.$$

Problem 235 *Suppose $u \in R^{n+1}$. For*

$$(\nabla_{S_n} \phi_n)(u)$$

given in Definition 20, show that

$$(\nabla_{S_n} \phi_n)(u) = (D^t D)^{-1} (\nabla \phi_n)(y),$$

where $(\nabla \phi_n)(u)$ is the conventional gradient of ϕ_n at u.

Problem 236 *Suppose that H is a Hilbert space, S is a closed subspace of H and $P \in L(H, H)$. Show that P is the orthogonal projection of H onto S if and only if the following four conditions hold:*

- *$P^2 = P$.*
- *$\langle Px, y \rangle_H = \langle x, Py \rangle_H$ for all $x, y \in H$.*
- *The range of P is a subset of S.*
- *If x is in the range of P, then $Px = x$.*

Problem 237 *Suppose P is the projection of $(R^n)^2$ onto the range of D. Show that*

$$P = D(D^t D)^{-1} D^t.$$

Problem 238 *Write a computer program which follows the iteration with the gradient in Problem 235. (Choose $n = 10, n = 20, n = 100$ and print your results.)*

Problem 239 *Make a comparison with the results from your code from Problem 238 with results from your code in Problem 232. Reflect further on the First Law of Numerical Analysis.*

Chapter 16
Semigroups and Families of Sobolev Spaces

Denote by H a Hilbert space. Denote by H' a Hilbert space whose points form a dense subset of H so that

$$\|x\|_{H'} \geq \|x\|_H, \ x \in H'.$$

Problem 240 *Show that*

$$H = L_2([0,1]), \ H' = H^{1,2}([0,1]) \qquad (16.1)$$

provides an example of the above setting (see [1] for definitions of the spaces in (16.1)) or see Definition 16 in Chapter 14.

Problem 241 *Returning to the general case (not just the example in Problem 240), suppose $y \in H$ and*

$$f(x) = \langle x, y \rangle_H, \ x \in H.$$

Denote by g the restriction of f to H'. Show that g is in $(H')^*$, the dual space of H'.

Problem 242 *Referring to either Problem 99 or 205, show that there is $z \in H'$ so that*

$$g(x) = \langle x, z \rangle_{H'}, \ x \in H'.$$

Denote by M the transformation so that if $y \in H$ as in Problem 241, then

$$My = z$$

where z is as in Problem 242.

M is called the embedding map between the spaces H and H'.

J.W. Neuberger, *A Sequence of Problems on Semigroups*,
Problem Books in Mathematics, DOI 10.1007/978-1-4614-0430-9_16,
© Springer Science+Business Media, LLC 2011

Problem 243 *Show that M above is linear.*

Problem 244 *Show that the range of M is dense in H.*

Problem 245 *Show that M^{-1} exists.*

Problem 246 *Show that*

$$\langle Mx, x \rangle_H > 0, \ x \in H, x \neq 0.$$

(That is, M is positive as a member of $L(H,H)$.)

Problem 247 *Show that*

$$|M|_{L(X,Y)} \leq 1$$

where

$$(X,Y) \text{ is any of the pairs } (H,H), (H,H'), (H',H), (H',H').$$

Problem 248 *Show that*

$$\langle x, My \rangle_H = \langle Mx, y \rangle_H, \ x, y \in H.$$

Problem 249 *Show that*

$$\langle x, My \rangle_{H'} = \langle Mx, y \rangle_{H'}, \ x, y \in H'.$$

Problem 250 *Show that M, considered as a member of $L(H,H)$, has a unique positive symmetric square root (denoted by $M^{1/2}$) See [66], for example.*

Problem 251 *Show that M, considered as a member of $L(H',H')$, has a unique positive symmetric square root and this square root agrees, on H', with the square root in Problem 250.*

Problem 252 *Show that if $x \in H'$, then*

$$\|x\|_H = \|M^{1/2}x\|_{H'}.$$

Problem 253 *Show that if $x \in H$, then*

$$\|x\|_H = \|M^{1/2}x\|_{H'}.$$

Problem 254 *Using Problem 102, and perhaps some material from the Notes to Chapter 7, show that if $\lambda \geq 0$, then there is a unique symmetric, positive transformation*

$$M^\lambda$$

so that if $\alpha, \beta \geq 0$, then

$$M^\alpha M^\beta = M^{\alpha+\beta}$$

and if $\alpha = \frac{m}{n}$, then

$$M^\alpha = (M^m)^{1/n}.$$

Problem 255 *Suppose that*

$$T(\lambda) = M^\lambda, \ \lambda \geq 0.$$

Show that T is a strongly continuous linear semigroup of contractions on H.

Problem 256 *Show that if $\lambda > 0$, then*

$$M^{-\lambda} \text{ exists}$$

(but perhaps is not continuous).

Problem 257 *If $\lambda > 0$, define*

$$H_\lambda = \text{ range of } M^{\lambda/2}$$

with

$$\|x\|_{H_\lambda} = \|M^{-\lambda/2}x\|_H, \ x \in H_\lambda.$$

Show that H_λ is a Hilbert space.

Problem 258 *For D as in Problem 235 show that*

$$(D^t D)^{-1} = M$$

where M is derived by the choice of

$$H = R^{n+1}, \ H' = S_n.$$

S_n is as in Problem 234.

Problem 259 *Consider M derived from H, H' as in Problem 240. Show that the usual norm (see [1]) for $H^{2,2}([0,1])$ is equivalent to the norm of H_2, as defined above for the pair (H, H').*

Problem 260 *Make a substantial generalization of Problem 259.*

Problem 261 *Consider practical numerical consequences of Problem 260. For example, let if D as in Problem 258 be replaced by a discrete version of a second-order derivative operator E. Compare the coding effort for the resulting $(E^t E)^{-1}$ with $(D^t D)^{-2}$.*

Problem 262 *Suppose that $n \in Z^+$, $C \in L(R^n, R^n)$ and*

$$\phi(x) = \frac{1}{2}\left\|\begin{pmatrix} x \\ Cx \end{pmatrix}\right\|^2, \ x \in R^n.$$

Find

$$\phi'(x), \ x \in R^n.$$

Problem 263 *For ϕ as in Problem 262, find the gradient of ϕ, that is, the function*

$$\nabla \phi : R^n \to R^n$$

such that

$$\phi'(x)h = \langle h, (\nabla \phi)(x) \rangle_{R^n}, \ x, h \in R^n.$$

Problem 264 *For ϕ as in Problem 263, find an explicit expression for $x \in R^n$ which minimizes $\phi(x)$.*

Problem 265 *Suppose that each of H, K is a Hilbert space and T is a closed, densely defined linear transformation on H into K. Show that the orthogonal projection onto*

$$\left\{ \begin{pmatrix} x \\ Tx \end{pmatrix} : x \in D(T) \right\}$$

is given by

$$P = \begin{pmatrix} (I + T^t T)^{-1} & T^t(I + TT^t)^{-1} \\ T(I + T^t T)^{-1} & I - (I + TT^t)^{-1} \end{pmatrix}, \tag{16.2}$$

a formula of von Neumann [70]. (Note Problems 103–107 in Chapter 7.)

Problem 266 *For T as in Problem 265 denote by H' the space whose points are exactly those of $D(T)$ with*

$$\|x\|_{H'} = \left\| \begin{pmatrix} x \\ Tx \end{pmatrix} \right\|_{H \times K}, \ x \in H'.$$

Show that

$$T \in L(H', K)$$

provided T is considered as a linear transformation from $H' \to K$.

Problem 267 *For T as in Problem 266, show that for $x \in H', y \in K$,*

$$\langle Tx, y \rangle_K = \langle \begin{pmatrix} x \\ Tx \end{pmatrix}, \begin{pmatrix} 0 \\ y \end{pmatrix} \rangle_{H \times K}$$

$$= \langle \begin{pmatrix} x \\ Tx \end{pmatrix}, P \begin{pmatrix} 0 \\ y \end{pmatrix} \rangle_{H \times K}.$$

Problem 268 *Following Problem 267 conclude that*

$$T^* y = \pi P \begin{pmatrix} 0 \\ y \end{pmatrix}, \ y \in K,$$

where P is as in Problem 265 and

$$\pi \begin{pmatrix} r \\ s \end{pmatrix} = r, \ r \in H, s \in K.$$

Also conclude that

$$\langle Tx, y \rangle_K = \langle x, T^* y \rangle_{H'}.$$

Problem 269 *For H, K, H' in Problem 266 define*

$$M = (I + T^t T)^{-1},$$

and show that M is the embedding operator for the pair (H, H').

Problem 270 *Review Chapter 14 discerning how developments of the present chapter can give more concrete expressions for the various Sobolev gradients in Chapter 14 and hence more concrete expressions for the semigroups of steepest descent there.*

Problem 271 *Take T, H, K, H' as in Problem 266. Write an essay organizing your thoughts on the following:*

With T a closed densely defined linear transformation on H to K, T has an adjoint T^t which is a closed densely defined linear transformation on K to H so that

$$\langle Tx, y \rangle_K = \langle x, T^t y \rangle_H, \ x \in D(T), y \in D(T^t).$$

On the other hand, this same transformation T, the same collection of ordered pairs, considered as a member of $L(H', K)$, T has an adjoint T^* so that

$$\langle Tx, y \rangle_K = \langle x, T^* y \rangle_{H'}, \ x \in H', y \in K.$$

In your essay, comment on how one transformation may have two adjoints, one everywhere discontinuous and the other continuous. Investigate how there may be a different adjoint for T corresponding to any space H'' with the same points as H' but with a norm $\| \cdot \|_{H''}$ different from that of H' but equivalent to it in the sense that there are $m, M > 0$ such that

$$m\|x\|_{H''} \le \|x\|_{H'} \le M\|x\|_{H''}, \ x \in H'.$$

Comment on how linguistic and notational ambiguities might present a hindrance to someone trying to deal with Sobolev gradients and matters of the present chapter. Can you think of a more rational notation that presents less of a hindrance?

Chapter 17
Nonlinear Semigroups Studied by Linear Methods

In this chapter, denote by X a separable complete metric space, that is to say, a Polish space.

Definition 22 *Suppose T is a semigroup on X. The statement that T is jointly continuous means that if*

$$g : [0, \infty) \times X \to X \qquad (17.1)$$

such that
$$g(t, x) = T(t)x, \ t \in [0, \infty), \ x \in X,$$

then g is continuous.

Definition 23 *$C(X)$ denotes the Banach space of all bounded and continuous functions $f : X \to R$. The norm in $C(X)$ is given by*

$$\|f\|_{C(X)} = \sup_{x \in X} |f(x)|.$$

Problem 272 *Suppose T is a jointly continuous semigroup on X and*

$$S : [0, \infty) \to \ \text{set of transformations on } C(X)$$

such that
$$(S(t)f)(x) = f(T(t)x), \ t \geq 0, x \in X, f \in C(X).$$

Show that if $t \geq 0$, then

$$S(t) \in L(C(X), C(X))$$

and
$$|S(t)| \leq 1.$$

S is called a representation of T.

Problem 273 *Show that S is a linear semigroup on $C(X)$.*

J.W. Neuberger, *A Sequence of Problems on Semigroups,*
Problem Books in Mathematics, DOI 10.1007/978-1-4614-0430-9_17,
© Springer Science+Business Media, LLC 2011

Problem 274 *Find an example of a semigroup S as in Problem 273 such that S is not strongly continuous under the sup norm of $C(X)$.*

Here is a second topology for $C(X)$. It is presented by means of a notion of convergence.

Definition 24 *We say that the sequence $f_1, f_2, \ldots$ in $C(X)$ β-converges to $f \in C(X)$ if*

$$\text{there exists } M > 0 \text{ such that } \|f_n\|_{C(X)} \leq M, \ n = 1, 2, \ldots,$$

and also $f_1, f_2, \ldots$ converges uniformly to f on each compact subset of X. In this case we write

$$\beta - \lim_{n \to \infty} f_n = f.$$

Problem 275 *Show that S in Problem 273 is β-strongly continuous in the sense that if*

$$f \in C(X) \quad \text{and} \quad \{t_n\}_{n=1}^{\infty} \text{ in } (0, \infty), \text{ converges to } 0,$$

then

$$\{S(t_n)f\}_{n=1}^{\infty} \ \beta\text{-converges to } f.$$

Problem 276 *Suppose $\{f_n\}_{n=1}^{\infty}$ is a sequence in $C(X)$, β-convergent to $f \in C(X)$, S as in Problem 273 and $t \geq 0$. Show that*

$$\{S(t)f_n\}_{n=1}^{\infty} \ \text{is } \beta\text{-convergent to } S(t)f.$$

Definition 25 *For T a jointly continuous semigroup on X, the Lie generator of T is*

$$A = \{(f, g) \in C(X)^2 : g(x) =$$

$$\lim_{t \to 0+} \frac{1}{t}(f(T(t)x) - f(x)), \ x \in X\}.$$

Problem 277 *Suppose that T is the semigroup on $[0, \infty)$ so that*

$$T(t)x = \frac{x}{1 + tx}, \ t \geq 0, x \geq 0.$$

Calculate the Lie generator A of T.

Problem 278 *Take additional examples of jointly continuous semigroups on a Banach space and calculate their Lie generators.*

Problem 279 *Show that the generator A in Definition 25 is a derivation in the sense that if $f, g \in D(A)$, then $fg \in D(A)$ and*

$$A(fg) = f(Ag) + (Af)g.$$

Definition 26 *For $\lambda > 0$ and T as in Problem 273, define a transformation I_λ with domain $C(X)$ such that*

$$(I_\lambda f)(x) = \frac{1}{\lambda} \int_0^\infty e^{-s/\lambda} f(T(s)x) \, ds, \; f \in C(X), \; x \in X.$$

Problem 280 *Show that if $\lambda > 0$, then $(I - \lambda A)I_\lambda = I$.*

Problem 281 *Show that if $\lambda > 0$ and f is in the domain of A, then*

$$I_\lambda (I - \lambda A)f = f.$$

Problem 282 *Show that if $\lambda > 0$,*

$$(I - \lambda A)^{-1}$$

exists and

$$(I - \lambda A)^{-1} = I_\lambda.$$

Problem 283 *Show that if λ and A are as in Problem 282, then*

$$\|(I - \lambda A)^{-1} f\|_{C(X)} \leq \|f\|_{C(X)}, \; f \in C(X).$$

Problem 284 *With A as in Problem 283 and $\{t_n\}_{n=1}^\infty$ a sequence in $(0, \infty)$ convergent to 0, show that if $n \in Z^+$, $\lambda \geq 0$ and $x \in X$, then*

$$((I - t_n A)^{-n} f)(x) = \int_0^\infty f(T(\cdot)x) \, d\phi_{\lambda,n}$$

where $\phi_{\lambda,n}$ is as in Problem 65.

Problem 285 *With A as in Problem 283 and $\{t_n\}_{n=1}^\infty$ a sequence in $(0, \infty)$ convergent to 0, show that*

$$\beta - \lim_{n \to \infty} (I - t_n A)^{-1} f = f,$$

and consequently, A has dense domain in the β-sense.

Problem 286 *With A and T given in this chapter and $\lambda > 0$, show that*

$$\beta - \lim_{n \to \infty} (I - \frac{\lambda}{n} A)^{-n} f = f(T(\lambda)), \; f \in C(X).$$

Definition 27 *A set Q of transformation from $C(X)$ to $C(X)$ is β-equicontinuous if for each sequence $\{f_n\}_{n=1}^\infty \in Q$, β-convergent to $f \in C(X)$, there is $M > 0$ such that*

$$\|W f_n\|_{C(X)} \leq M, \; W \in Q, \; n = 1, 2, \ldots,$$

and if Ω is a compact subset of X and $\epsilon > 0$, there is N such that if $n > N$,

$$|(Wf_n)(x) - (Wf)(x)| < \epsilon, \ x \in \Omega, \ W \in Q.$$

Problem 287 *Show that if $\eta > 0$, then*

$$\{(I - \lambda/nA)^{-n}, \ 0 \le \lambda \le \eta, \ n = 1, 2, \dots\}$$

is β-equi-continuous.

Definition 28 *Denote by $LG(X)$ the set of all linear transformations A with domain and range in $C(X)$ which have the properties given in Problems 279, 283, 285, 287, that is,*

- *A is a derivation.*
- *The domain of A is dense in $C(X)$ in the β-sense.*
- *If $\lambda \ge 0$, $(I - \lambda A)^{-1}$ exists and is a contraction in $L(C(X), C(X))$.*
- *If $\eta \ge 0$, $\{(I - (\lambda/n)A)^{-n}, \ 0 \le \lambda \le \eta, n = 1, 2, \dots\}$ is β-equi-continuous.*

Denote by $LG(X)$ the collection of all such transformations A satisfying the above items.

Problem 288 *Make a theorem from Problems 273–287.*

For Problems 289–292 suppose that $A \in LG(X)$. Define

$$I_\lambda = (I - \lambda A)^{-1} \ \text{for } \lambda > 0.$$

Problem 289 *Show that there is a linear semigroup $\overline{V}$, which is strongly continuous (in norm of $C(X)$ but on $\overline{D(A)}$), such that*

$$\overline{V}(\lambda)f \ = \ \lim_{n \to \infty} (I_{\lambda/n})^{-n}f, \ \lambda > 0, f \in D(A).$$

Problem 290 *Show that there is a unique β-continuous extension S of $\overline{V}$, to all of $C(X)$.*

Problem 291 *Show that S in Problem 290 is a β-continuous linear semigroup on all of $C(X)$.*

Problem 292 *Show that if S as in Problem 290, then*

$$A \ = \ \{(f, g) \in C(X) : g \ = \ \beta - \lim_{t \to \infty} \frac{1}{t}(S(t)f - f)\}.$$

Problem 293 *Suppose that $z \in X$ and $\mu : C(X) \to R$ is such that*

$$\mu(f) = f(z), \ f \in C(X).$$

Show that μ is linear, β-continuous and that

$$\mu(f)\mu(g) = \mu(fg), \ f, g \in C(X),$$

i.e., that μ is a β-continuous multiplicative linear functional on $C(X)$.

Problem 294 *For S as in Problem 290, $t \geq 0$, $x \in X$ and μ with domain $C(X)$ such that*

$$\mu(f) = (S(t)f)(x), \ f \in C(X),$$

show that μ is a transformation as in Problem 293.

Problem 295 *Show that if μ is a nonzero β-continuous multiplicative linear functional on $C(X)$, then there is a unique point $z \in X$ such that*

$$\mu(f) = f(z) \text{ for all } f \in C(X).$$

See [14], in particular the result there due to J. Lawson.

Problem 296 *Define T with domain $[0, \infty)$ so that if $t \geq 0$, then $T(t)$ is the transformation with domain X and range in X such that if $x \in X$, then $T(t)x$ is the element $z \in X$ such that*

$$(S(t)f)(x) = f(z), \ f \in C(X).$$

Show that T is a semigroup on X.

Problem 297 *Show that T in Problem 296 is jointly continuous.*

Problem 298 *Show that the Lie generator A of T satisfies the equation in Definition 25.*

Problem 299 *Write an essay summarizing results of this chapter.*

Chapter 18
Measures and Linear Extension of Nonlinear Semigroups

Notation from Chapter 17 holds and the following is added:

Definition 29 *Denote by $M(X)$ the set of all Borel measures μ on X and by $B(X)$ the set of all Borel subsets of X.*

Problem 300 *Show that $M(X)$ in Definition 29 is a representation of the second dual of $C(X)$ in Chapter 17. See [67] for more information.*

Definition 30 *Suppose*

$$\mu \in M(X) \text{ and its range is a subset of } [0, \infty).$$

We say that μ is compact regular if

$$\mu(\Omega) = \sup\{\mu(\Omega') : \Omega' \subset \Omega \text{ and } \Omega' \text{ is compact}\}.$$

Definition 31 *More generally, we also say that $\mu \in M(X)$ is compact regular if μ is the difference of two compact regular measures as in Definition 30 which have range in $[0, \infty)$. In this case we say that $\mu \in MCR(X)$.*

Definition 32 *Suppose T is a jointly continuous semigroup on X. We say that U extends T if U is a function with domain $[0, \infty)$ and*

$$(U(t)\mu)(\Omega) = \mu\{T(t)^{-1}\Omega\}, \ t \geq 0,$$

with $\Omega \in B(X), \mu \in MCR(X)$.

Problem 301 *Show that U in Definition 32 is a semigroup on $MCR(X)$.*

Problem 302 *Show that U in Definition 32 is a linear semigroup on $MCR(X)$.*

J.W. Neuberger, *A Sequence of Problems on Semigroups*,
Problem Books in Mathematics, DOI 10.1007/978-1-4614-0430-9_18,
© Springer Science+Business Media, LLC 2011

Definition 33 *For U in Definition 32, define*

$$C = \{(\mu, \nu) \in MCR(X)^2 : \int_X f \, d\nu =$$

$$\lim_{t \to 0+} \int_X \frac{1}{t}(f(T(t)) - f) \, d\mu, \ f \in C(X)\}.$$

(The integrals are Lebesgue integrals.) We say that C is the extended generator of T.

Definition 34 *Denote by $C(X)^{*\beta}$ the linear space of all β- continuous linear functions g,*

$$g : C(X) \to R,$$

such that g is continuous in the β- sense.

Problem 303 *Suppose $\mu \in MCR(X)$ and $p : C(X) \to R$ such that*

$$pf = \int_X f \, d\mu, \ f \in C(X).$$

Show that

$$p \in C(X)^{*\beta}.$$

Problem 304 *Suppose $p \in C(X)^{*\beta}$. Show that there is $\mu \in MCR(X)$ such that*

$$pf = \int_X f \, d\mu, \ f \in C(X).$$

(It might be helpful, for this problem, to find some lemmas in [67].)

Problem 305 *For C given in Definition 33, show that if $\lambda > 0$, then*

$$(I - \lambda C)^{-1} \ exists$$

with domain all of $MCR(X)$ and

$$|\int_X f \, d((I - \lambda C)^{-1}\mu)| \le |f| \, |\int_X |d\mu|, \ f \in C(X).$$

Problem 306 *For C as in Definition 33, show that if $\lambda \ge 0$, then*

$$\lim_{n \to \infty} \int_X f \, d((I - \lambda C)^{-n}\mu) = \int_X f \, d(U(\lambda)\mu), \ f \in CB(X).$$

Problem 307 *For C in Definition 33, show that there exists $A \in LG(X)$ such that*

$$\int_X f \, d(C\mu) = \int_X Af \, d\mu. \tag{18.1}$$

Problem 308 *Suppose $A \in LG(X)$ (see Definition 28) and C a linear transformation*

$$MCR(X) \to MCR(X)$$

so that (18.1) is satisfied. Show that there exists a unique semigroup on X that has C as its extended generator.

Problem 309 *Make a theorem which summarizes Problems 301–308.*

Problem 310 *Generalize results of the previous chapter to include the setting of the present chapter.*

Problem 311 *Refer to Problem 14 for a definition of a certain semigroup T on $X = C([-1,1])$. Review your work on these two problems. Try to conclude that the domain of the conventional generator of T contains only functions that are nonnegative or else are negative. Contemplate difficulties in using such a generator for purposes of recovering T in terms of it. Is C of Definition 33 a more promising choice of generator than that from Problem 14? See Notes to the present chapter to further understand the importance of this example.*

Problem 312 *For T as in Problem 14, calculate a generator A for T as in Chapter 17. In particular, determine S on $C(X)$ so that*

$$(S(t)f)(x) = f(T(t)x), \ t \geq 0, \ x \in X,$$

being as concrete as possible. Try to make an analysis of T in terms of A.

Problem 313 *Relate C of Problem 311 to A of Problem 312.*

Chapter 19
Local Semigroups and Lie Generators

Some nonlinear differential equations do not generate a semigroup such as those that are the subject of previous problems. For example there is:

Problem 314 *Investigate solutions u to*

$$u(0) = x \geq 0, \ u' = u^2 \tag{19.1}$$

where for each $x \geq 0$, the interval of existence contains zero, is a subset of $[0, \infty)$ and is as long as possible.

Problem 315 *For the setting of Problem 314, define T so that*

- *If $t \geq 0$, then $T(t)$ is a function from a subset of $[0, \infty)$ into $[0, \infty)$.*
- *If $x \geq 0$, and x is in the domain of $T(t)$ if and only if there is a solution u to (19.1) so that t is in the domain of u satisfying (19.1).*
- *If $x \geq 0$, then $T(t)x = u(t)$ where u satisfies (19.1).*

Articulate as many properties of T as you can.

Definition 35 *Suppose that X is a complete separable metric space. The statement that T is a local semigroup on X means that*

- *There is ω so that $0 < \omega \leq \infty$.*
- *If $t < \omega$, then $T(t)$ is a function from a subset of X into X.*
- *$T(0)x = x$, $x \in X$.*
- *There is a function $m : X \to (0, \infty]$ such that $1/m$ is continuous and so that $x \in D(T(t)) \iff t \in [0, m(x))$.*
- *$m(x) < \infty$ for some $x \in X$.*
- *If $t, s \geq 0$ and $x \in X$, then*

$$T(t)T(s)x = T(t+s)x \iff t + s < m(x).$$

- *T is jointly continuous.*
- *T is maximal, i.e., if $s \in [0, \infty)$ and $\lim_{t \to s-} T(t)x$ exists, then $s < m(x)$.*

J.W. Neuberger, *A Sequence of Problems on Semigroups*,
Problem Books in Mathematics, DOI 10.1007/978-1-4614-0430-9_19,
© Springer Science+Business Media, LLC 2011

Problem 316 *Show that T from Problem 315 is a local semigroup.*

Definition 36 *Suppose that T is a local semigroup on X. Then the Lie generator of T is*

$$A = \{(f,g) \in C(X)^2 : g(x) = \lim_{t \to 0+} \frac{1}{t}(f(T(t)x) - f(x)),\ x \in X.$$

Problem 317 *Calculate the Lie generator for T in Problem 315.*

For Problems 318 to 323, suppose that T is a local semigroup on X and A is its Lie generator.

Problem 318 *Show that A is a linear derivation, i.e., if $f, g \in D(A)$, then $fg \in D(A)$ and*
$$A(fg) = f(Ag) + (Af)g.$$

Problem 319 *Show that $D(A)$ is β-dense in $C(X)$.*

Problem 320 *Suppose that $\lambda > 0$. Define I_λ on $C(X)$ so that if $f \in C(X)$, then*
$$(I_\lambda f)(x) = \frac{1}{\lambda} \int_0^{m(x)} \exp(-j/\lambda) f(T(j)x),\ x \in X.$$

Show that

- *I_λ is a linear transformation from $C(X)$ to $C(X)$.*
- *$|I_\lambda| \leq 1$.*
- *$(I - \lambda A)I_\lambda f = f,\ f \in C(X)$.*

Problem 321 *Use assumptions and notation of Problem 320. Show that if $f \in D(A)$, then*
$$I_\lambda(I - \lambda A)f = f - g$$
for $g \in C(X)$ for

$$g(x) = \lim_{t \to m(x)-} \exp(-t/\lambda) f(T(t)x),\ x \in X.$$

Problem 322 *Suppose that g, λ, T, A, f are as in Problem 321. Show that*

$$(I - \lambda A)g = 0.$$

Show that if $f(y) > 0, y \in X$, then the resulting member $g \in C(X)$ is an eigenvector of A, if g is not zero.

Problem 323 *Using assumptions and notation of Problem 320, show that*

$$\lim_{n \to \infty} (I_{\lambda/n}^n f)(x) = f(T(\lambda)x),\ f \in C(X),\ x \in X.$$

Problem 324 *For T as in Problem 315, verify all the conclusions to Problems 318–323.*

Problem 325 *For T as in Problem 315 calculate the corresponding element g as in Problem 321 for various $f \in C(X)$.*

Problem 326 *Suppose T is a local semigroup. Define m as in Definition 35 and that*

$$f(x) = \exp^{-m(x)}, \ x \in X.$$

Show that

$$Af = f.$$

Problem 327 *For f as in Problem 326, show that*

$$A(f^2) = 2f.$$

Problem 328 *Generalize the result of Problem 327.*

Problem 329 *Suppose T is a local semigroup and $\lambda > 0$. Define m as in Definition 35 and*

$$f(x) = \exp^{-\lambda m(x)}, \ x \in X.$$

Then

$$Af = \lambda f.$$

Problem 330 *Show that the following is true: Suppose that T is either a local semigroup or a global semigroup and A is its Lie generator. Then T is global if and only if A has no positive eigenvalue.*

Problem 331 *Pick various systems of time-dependent autonomous differential, ordinary or partial differential equations that are known to generate either a local or global semigroup. For such systems, calculate a Lie generator. Use some of the results of the problems in this chapter to try to decide if corresponding semigroups are local or global. The problem of three-dimensional Navier–Stokes is such a problem, surely not an easy one.*

Problem 332 *Dream of a numerical attack corresponding to Problems 330 and 331. For various choices of X and T, make a discrete version of X, then a discrete version of the Lie generator A of T on a discrete version of C(X). Then use numerics in order to test whether this discrete version of A might have a positive eigenvalue. Consider first testing for situations for which the answer is known.*

The following problems are late additions to the problem set. The 'dream' mentioned in Problem 332 has come a little closer to being realized. Within the past year, I have written a code in which X is one-dimensional, i.e., to test a single ordinary ODE for global or local existence in time. J. W. Swift and John M. Neuberger have written a code for cases in which X is two-dimensional. Preliminary results are encouraging.

Problem 333 *Write a MatLab code to test various one-dimensional examples for global versus local existence. Examples might include*

- $u' - u^2 = 0$, $X = [0, b)$, $0 < b \leq \infty$.
- $u' - u(1 - u) = 0$, $X = [0, b)$, $0 < b \leq \infty$.
- $u' + u^2 = 0$, $X = [0, b)$, $0 < b \leq \infty$.

The eigenvalue–eigenvector routines from MatLab may be used.

Problem 334 *Consider the use of finite element approximations, first in one-, then two-dimensional problems in connection with Problem 333. Realize that in the present context, 'two-dimensional' problems refer to a system of two ordinary differential equations.*

Problem 335 *Consider the local–global existence problem for the partial differential equation which seeks, for some $w > 0$, a function u with domain $[0, w] \times [0, 1]$ so that*

$$u_1(t, x) = mu_{2,2}(t, x) + \epsilon |u(t, x)|^p \tag{19.2}$$

for some $p > 1$ and some $m > 0$. For boundary conditions, take

$$u(t, 0) = 0 = u(t, 1), \ t \in [0, w],$$

and

$$u(0, x) = f(x), \ x \in [0, 1],$$

for some given function f. Set up a corresponding finite-dimensional version and investigate the existence or nonexistence of eigenvectors of the relevant Lie generator in order to decide which of local or global describes the underlying semigroup.

Problem 336 *Realize that for Equation 19.2, for $\epsilon = 0$ this is just the heat equation, which has existence on $[0, w]$ for any $w > 0$ for reasonable f. Realize also that for $m = 0$, Equation 19.2 is essentially an uncountable system of ordinary differential equations. Consider that in a proper discretization of this problem the number of grid points will be very large. Estimate how many grid points it would take for such a proper discretization if results of this chapter are to be applied.*

Problem 337 *Now back to dreaming. Write down a time-dependent Navier–Stokes system in three space dimension. Make an estimate of the number of grid points needed for a reasonable finite-dimensional version of a transformation A for which a positive eigenvalue indicates only local existence and for which the lack of such an eigenvalue is indicative of global existence. Try to form an opinion as to whether any present-day computer is sufficient for this task. If your opinion is negative on this matter, is that sufficient reason for not working on this problem? Do you believe that the power of computers will increase in the future somewhat as they have in the past?*

Another instance where 'local' versus 'global' solutions is of interest is the following:

Problem 338 *Suppose that n is a positive integer, F is a polynomial function: $C^n \to C^n$ so that for each $x \in C^n$, the derivative $F'(x)$ has an inverse. Is it true that for each $x \in C^n$ there is $z : [0,1] \to C^n$ so that*

$$z(0) = x, \ z'(t) = -(F'(z(t))^{-1}F(x), \ t \in [0,1]?$$

Problem 339 *Show that an affirmative answer, for each such function F in Problem 338, implies the truth of:*

Jacobian Conjecture: Every F as in Problem 338 is a bijection.

See [72] and [50] for some background on the Jacobian Conjecture. A Web search yields an extensive literature on this problem. The problem is unsolved by anyone, if $n > 1$, so far as I know.

Chapter 20
Boundary (Supplementary) Conditions for Partial Differential Equations

We first ask for some essays to help delineate a certain point of view.

Problem 340 *Write an essay on your view of how the problem of finding roots of polynomials developed over the ages. To help you get started, note the Babylonian's progress on solving quadratic equations.*

Problem 341 *Write an essay on your view of how the problem of solving systems of ordinary differential equations has developed over time. Perhaps comment first on the early role of 'closed form' solutions.*

Problem 342 *Write an essay on your view of how the problem of solving systems of partial differential equations has developed over time.*

Problem 343 *Contrast the points of view at the present time on the subject matters of Provlems 340,341,342.*

Problem 344 *Comment on how the advent of modern computers may have helped change the points of view on each of Problems 340, 341, 342.*

Problem 345 *Show that if $f : R \to R$ is a C^1 function, then there is a unique function*

$$u : R^2 \to R$$

such that

$$u_1 + u_2 = 0 \text{ and } u(0, y) = f(y), \ y \in R.$$

Problem 346 *Do you agree that your results on Problem 345 gives a characterization of the set of all solutions to the partial differential equation in that problem?*

Problem 347 *Given a single linear second order constant coefficient ordinary differential equation on the interval $[0, 1]$, show that the set of all solutions has a characterization as a two dimensional subspace of $C([0,1])$, that is, the set of all solutions is one-to-one with R^2.*

J.W. Neuberger, *A Sequence of Problems on Semigroups*,
Problem Books in Mathematics, DOI 10.1007/978-1-4614-0430-9_20,
© Springer Science+Business Media, LLC 2011

Problem 348 *After, perhaps, reconsidering Chapter 14, indicate your opinion of how the finding of critical points of a real-valued function ϕ on a Hilbert space H relates to the solving of systems of partial differential equations.*

Standing Hypothesis for Problems 349–354:

Suppose H is a Hilbert space, $\phi : H \to [0, \infty)$ is a C^1 function and $\nabla \phi$ is the function on H so that

-

$$\phi'(x)h = \langle h, (\nabla \varphi)(x) \rangle_H, \; u, h \in H.$$

- $\nabla \phi$ is locally lipschitzian.
- $G : H \to H$ is a continuous function such that if $x \in H$ and

$$z(0) = x, \; z'(t) = -\nabla \phi(z(t)), \; t \geq 0, \tag{20.1}$$

then $G(x) = \lim_{t \to \infty} z(t)$.
- $C(H)$ is the Banach space, under sup norm of all bounded continuous real-valued functions on H.
- T is the semigroup generated by (20.1).
- A is the Lie generator of T.

Problem 349 *Show that if $x \in H$, then*

$$G(T(\cdot)x) \text{ is constant.}$$

Problem 350 *Show that if $x, y \in H$, there is $f \in C(H)$ such that $f(x) \neq f(y)$.*

Problem 351 *Show that if $f \in N(A)$, the null space of A, and $x \in H$, then*

$$(Af)(T(\cdot)x) \text{ is constant.}$$

Problem 352 *Show that if $x \in H$ and*

$$y \in \bigcap_{g \in N(A)} g^{-1}(g(x)),$$

then

$$y \in G^{-1}(G(x)).$$

Problem 353 *Show that if $x \in H$ and*

$$y \in G^{-1}(G(x)),$$

then

$$y \in \bigcap_{g \in N(A)} g^{-1}(g(x)).$$

Problem 354 *Show that each equivalence class*

$$\{G^{-1}(G(x)), x \in H\}$$

contains precisely one $u \in H$ such that

$$\nabla \phi(u) = 0.$$

Problem 355 *Suppose that $H = H^{1,2}([0,1])$ and*

$$\phi(u) = \frac{1}{2}\|u' - u\|^2, \ u \in H.$$

Determine the equivalence classes

$$\{G^{-1}(G(x)), x \in H\} \tag{20.2}$$

associated with ϕ as in Problem 354.

Problem 356 *Suppose that*

$$\phi(u) = \frac{1}{2}\|u' + u^2\|^2.$$

Try to determine the equivalence classes

$$G^{-1}(G(x)), x \in H$$

for this choice of ϕ.

Problem 357 *Determine the equivalence classes associated with the equation in Problem 345.*

Problem 358 *Think about the following: For collections of problems based on a function ϕ satisfying the standing hypothesis, Problems 352 and 353 give a set of objects (equivalence classes) each of which contains one and only one critical point of ϕ. Granted the nature of a corresponding Lie generator A is understood, then this set of equivalence classes gives a more or less constructive characterization of the set of all critical points of ϕ. How do such characterizations differ from the more common ones which attempt to associate unique solutions with conditions given on a boundary of the region on which a problem is defined?*

Problem 359 *Keep an open mind about finding alternatives to developments of this chapters—developments which associate solutions with constructively given sets.*

Chapter 21
Quasianalyticity and Semigroups

This chapter contains a development for linear semigroups which is of interest in the theory of probability and other areas.

Definition 37 *Suppose f is a real-valued function and $u, \delta \in R$. If $n \in Z^+$ and*

$$[u, u + \delta n] \subset \quad domain \ of \ f,$$

denote the difference of order n for f over the interval $[u, u + \delta n]$ by

$$\Delta_f(n; u, \delta) = \Sigma_{k=0}^n \binom{n}{k}(-1)^{n-k} f(u + k\delta).$$

Problem 360 *Suppose that $[a, b]$ is an interval, $n \in Z^+$ and f is a real-valued C^n function on $[a, b]$. Show that there is $c \in (a, b)$ so that*

$$\lim_{\delta \to 0} \frac{\Delta_f(n; u, \delta)}{\delta^n} = f^{(n)}(c).$$

Problem 361 *Suppose f is a real-valued function, which is bounded by $M \geq 0$, whose domain contains the interval $[a, b]$. Show that if*

$$n \in Z^+, u, \delta \in R \ and \ [u, u + \delta n] \subset domain \ of \ f,$$

it follows that

$$|\Delta_f(n; u, \delta)| \leq M 2^n.$$

The next problem is a result of a theorem of Arne Beurling [4], p. 429. It is recommended that the reader consult this reference.

Problem 362 *Suppose f is a continuous function on $[a, b]$ for which there are $M, \epsilon > 0$ such that*

$$|\Delta_f(n; u, \delta)| \leq M(2 - \epsilon)^n$$

J.W. Neuberger, *A Sequence of Problems on Semigroups*,
Problem Books in Mathematics, DOI 10.1007/978-1-4614-0430-9_21,
© Springer Science+Business Media, LLC 2011

if $[u, u + \delta n]$ is a subset of the domain of f. Show that f is real-analytic at all $x \in (a, b)$.

Definition 38 *Suppose G is a collection of real-valued functions which have $[a, b]$ as their common domain. We say that G is quasianalytic if no two elements of G agree on any nondegenerate subinterval of $[a, b]$.*

The next two problems are results that appear in [42], which an interested reader is encouraged to consult.

Definition 39 *A continuous function f on $[0, 1]$ is called* unpredictable *provided that*

- $f(x) = 0$ *if $0 \le x \le \frac{1}{2}$,*
- *if $\epsilon > 0$ there is $x \in (\frac{1}{2}, \frac{1}{2} + \epsilon)$ such that $f(x) \ne 0$.*

Problem 363 *Suppose f is an unpredictable function and $\frac{1}{2} \le a < b < 1$. Show that*

$$\lim_{\delta \to 0+} \frac{\delta}{b - a} \sum_{n\delta \in [a,b]} \left(\frac{|\Delta_f(n; 0, \delta)|}{H(n, \delta)} \right)^{1/n} = 1 \qquad (21.1)$$

in which

$$H(n, \delta) = C(n, k)$$

where k is the least member of Z^+ so that $\frac{k}{n} > \frac{1}{2}$. See [42] for background. This is the hardest problem in the present book, in my opinion. Actually, it is the hardest problem I know how to solve.

Problem 364 *Suppose*

$$\{\delta_k\}_{k=1}^{\infty}$$

is a decreasing sequence convergent to zero and G is a collection of real functions with common domain $[a, b]$ such that if $f \in G$, there is $M, \epsilon > 0$ such that

$$|\Delta_f(n; u, \delta_k)| \le M(2 - \epsilon)^n \text{ if } [u, u + \delta_k n] \subset [a, b].$$

Show that G is a quasianalytic collection.

Problem 365 *Suppose T is a strongly continuous linear semigroup of contractions on the Banach space X. Show that if $x \in X$, $f \in X^*$ and*

$$g(t) = f(T(t)x), \ t \ge 0, \qquad (21.2)$$

then

$$|\Delta_g(n; u, \delta)| \le |f| \|x\| |T(\delta) - I|^n.$$

One says that the function g in (21.2) is a functional of a trajectory (an 'fot') of T.

Problem 366 *Suppose T is a linear strongly continuous semigroup of contractions on the Banach space H and*

$$\lim_{t \to 0+} \sup |T(t) - I| < 2. \qquad (21.3)$$

Show that each 'fot' of T is real analytic at all $t > 0$.

Problem 367 *Suppose T is a strongly continuous semigroup of contractions on the Banach space H and*

$$\lim_{t \to 0+} \inf |T(t) - I| < 2. \qquad (21.4)$$

Show that the set of all 'fot' of T is a quasianalytic collection.

Problem 368 *Suppose that each of $\{p_{i,j}\}_{i,j=0,1,\dots}$ is a continuous function on $[0, \infty)$ into $[0, 1]$ so that*

$$p_{i,j}(t + s) = \sum_{k=0}^{\infty} p_{i,k}(t) p_{k,j}(s), \ t, s \geq 0$$

and

$$\sum_{j=0}^{\infty} p_{i,j}(t) = 1, \ t \geq 0, \ i = 0, 1, \dots.$$

Define P as a function with domain $[0, \infty)$ so that if

$$x = \begin{pmatrix} x_0 \\ x_1 \\ \vdots \end{pmatrix}$$

is in ℓ_1, then

$$P(t)x = \begin{pmatrix} p_{0,0}(t) & p_{0,1}(t) & \cdots \\ p_{1,0}(t) & p_{1,1}(t) & \cdots \\ \vdots & \vdots & \end{pmatrix} \begin{pmatrix} x_0 \\ x_1 \\ \vdots \end{pmatrix} \ t \geq 0.$$

Furthermore, denote by g the function on $[0, \infty)$ so that

$$g(t) = \inf_{i=0,1,\dots} p_{i,i}(t), \ t \geq 0.$$

Show that

$$|P(t) - I| = 2(1 - g(t)), \ t \geq 0,$$

where the indicated norm is for members of $L(\ell_1.\ell_1)$.

P is called a denumerable Markov semigroup and members of $\{p_{i,j}\}_{i,j \geq 0}^{\infty}$ are transition probabilities for P, $p_{i,j}(t)$ representing the probability of transitioning from state i to state j in time t.

Problem 369 *Show that P in Problem 368 is a strongly continuous semigroup on ℓ_1.*

Problem 370 *Using the setting of Problem 368, if each of i and j is a non-negative integer, find $f \in \ell_1^*$ and $x \in \ell_1$ so that*

$$p_{i,j}(t) = f(P(t)x),\ t \geq 0.$$

Problem 371 *Give a probabilistic interpretation of the function g in Problem 368. In particular, interpret, in terms of probability, what it means for*

$$\lim_{t \to 0+} \sup g(t) = 0.$$

Problem 372 *Give a probabilistic interpretation of what it means for*

$$\lim_{t \to 0+} \inf g(t) = 0.$$

Problem 373 *Give a probabilistic interpretation of what it means for*

$$\lim_{t \to 0+} \inf g(t) < 2.$$

Problem 374 *Give a probabilistic interpretation of what it means for*

$$\lim_{t \to 0+} \sup g(t) < 2.$$

Problem 375 *Apply results of Problems 366 and 367 to Problem 368 using Definition 38. See [26] for more information.*

Problem 376 *For the semigroup T in Problem 94, show that*

$$\lim_{t \to 0+} \sup |T(t) - I| < 2. \tag{21.5}$$

Problem 377 *Improve on the bound in (21.5).*

Problem 378 *For T in Problem 94 and A its generator, show that*

$$AT(t) \text{ is a continuous linear transformation for each } t \geq 0.$$

Problem 379 *Show that if T is as in Problem 366 and A is its generator, then*
$$AT(t) \text{ is a continuous linear transformation, } t \geq 0.$$

Problem 380 *Suppose that T is a strongly continuous linear semigroup of linear transformations on the Banach space X such that*

$$\lim_{t \to 0+} |T(t) - I| < 2$$

and that T may be extended to a strongly continuous group of members of $L(X, X)$ on all of R. Show that it follows that T is continuous and the generator A of T is also continuous.

Problem 381 Show that if T is as in Problem 378, then T may be extended to the right half complex plane $C+$ so that it remains a semigroup on that set and that moreover,

$$T(\cdot) \text{ is an analytic function from } C+ \to L(X, X).$$

Problem 382 Suppose X is a Banach space, T is a strongly continuous semigroup on X and r, s, ρ are positive numbers with $\rho \in (1, 2)$, $s < r$. Suppose also that $M > 0$ and that if $n \in Z^+$, $h > 0$, then

$$|(Tf(h) - I)^n| < M\rho^n \text{ if } nh \in [r, s].$$

Show that there is $b > 0$ such that if $x \in X$, then the function

$$g: g(t) = T(t)x, \ t > b$$

is analytic. (See [11] for background and more information.)

Problem 383 For this problem by ℓ_2 is meant complex ℓ_2. Define T so that if

$$x = \{x_1, x_2, \dots\} \in \ell_2,$$
$$T(t)x = \{x_1 e^{it}, x_2 e^{2it}, x_3 e^{3it}, \dots\}, \ t \in R.$$

Show that T is a strongly continuous group of linear transformations on ℓ_2.

Problem 384 Determine a generator for T in Problem 383.

Problem 385 For T as in Problem 383, show that

$$\lim_{t \to 0} |T(t) - I| = 2.$$

Problem 386 For T as in Problem 383 find two 'fot' for T which agree on some open subset of $[0, \infty)$.

Problem 387 Suppose that $m, n \in Z^+$, $\delta > 0$ and

$$g(x) = e^{imx}, \ x \in R.$$

Show that

$$|\Delta_g(n, u, \delta)| \leq 2^n |\sin(\frac{m\delta}{2})|^n.$$

Problem 388 Suppose that

$$\{a_m\}_{m=0}^{\infty}$$

is a sequence in C such that

$$\sum_{m=0}^{\infty} |a_m| < \infty.$$

Suppose also that

$$h(x) = \sum_{m=0}^{\infty} a_m e^{imx}, \ x \in R.$$

For $\delta > 0$, find an inequality for

$$|\Delta_h(n; u, \delta)|$$

using the results of Problem 387.

Problem 389 *Suppose $\delta > 0$. What conditions on $\{a_m\}_{m=0}^{\infty}$ in Problem 388 lead to an inequality, for some $M, \epsilon > 0$,*

$$|\Delta_h(n; u, \delta)| \leq M(2 - \epsilon)^n,$$

for all $n \in Z^+, u \in R, \delta \in R$?

Problem 390 *Find an increasing sequence $\{n_k\}_{k=1}^{\infty}$ so that if T is the group given by*

$$T(t)x = \{x_{n_1} e^{in_1 t}, x_{n_2} e^{in_2 t}, x_{n_3} e^{in_3 t}, \dots \}, \ t \in R$$

for all

$$x = \{x_{n_k}\}_{k=1}^{\infty} \in \ell_2,$$

then

$$\liminf_{t \to 0} |T(t) - I| < 2.$$

Problem 391 *Find a quasianalytic collection G as in Problem 364 so that every 'fot' (as in Problem 365) of T is in G.*

Problem 392 *Attempt to develop a generalization of results of this chapter to nonlinear semigroups. Consider the development in Chapter 17. Does that help?*

Chapter 22
Continuous Newton's Method and Semigroups

Suppose that X is a Banach space and $F : X \to X$ is a C^2 function so that $(F'(x))^{-1}$ exists and is in $L(X, X)$ for all $x \in X$. Suppose also that $g \in X$.

Problem 393 *Show that, given $g \in X$, there is $0 < c \le \infty$ for which there is a unique function $u : [0, c)$ so that*

$$u(0) = g \text{ and } u'(t) = -(F'(u(t)))^{-1} F(u(t)), \ t \in [0, c). \tag{22.1}$$

Problem 394 *For u, c, F, X, g as in Problem 393, show that*

$$F(u(t)) = \exp(-t) F(g), \ t \in [0, c).$$

Problem 395 *Suppose $c = \infty$ in Problem 393. Show that (22.1) can be rescaled to $[0, 1)$ to be equivalent to finding v so that*

$$v(0) = g, \ v'(t) = -(F'(v(t)))^{-1} F(g), \ t \in [0, 1).$$

Problem 396 *Show that if $c = \infty$ in Problem 393 and*

$$w = \lim_{t \to \infty} u(t)$$

exists, then
$$F(w) = 0.$$

Problem 397 *Show that if v in Problem 395 can be extended by continuity to $[0, 1]$ (i.e., that*

$$v(1) = \lim_{t \to 1} v(t)$$

exists), then
$$F(v(1)) = 0.$$

Problem 398 *Suppose that each of m and n is a positive integer. Suppose furthermore that $r > 0$, $x \in R^n$ and that F is a continuous function from*

J.W. Neuberger, *A Sequence of Problems on Semigroups*,
Problem Books in Mathematics, DOI 10.1007/978-1-4614-0430-9_22,
© Springer Science+Business Media, LLC 2011

$\{z \in R^n : \|z - x\| \leq r\}$ to R^m with the property that if $\|y - x\| < r$, then there is $h \in R^n$ so that $\|h\| < r$ and

$$\lim_{t \to 0+} \frac{1}{t}(F(y + th) - F(y)) = -F(x).$$

Show that there is $u \in R^n$ so that $\|u - x\| \leq r$ and $F(u) = 0$.

A problem preliminary to Problem 398 follows:

Problem 399 *Suppose $\epsilon > 0$. Under the assumptions of Problem 398, define*

$$S = \{s \in [0, 1] :$$
$$\exists y \in R^n, \|y - x\| \leq rs, \|F(y) - (1 - s)F(x)\| \leq \epsilon s\}.$$

Show that S is closed and that $\sup S = 1$.

Definition 40 *Suppose that each of H and J is a Banach space. The statement that H is compactly embedded in J means that*

- *The points of H form a dense subspace of J.*
- *If $y_1, y_2, \ldots$ is a sequence in H so that for some $M > 0$, $\|y_k\|_H \leq M$, then $y_1, y_2, \ldots$ has a subsequence convergent in J to an element $y \in H$ such that $\|y\|_H \leq M$.*

Problem 400 *Suppose H is the space $H^{1,2}([0,1])$ as in Definition 16 and $J = C([0,1])$, the Banach space of all continuous real-valued functions on $[0,1]$ with*

$$\|f\|_J = sup_{s \in [0,1]}|f(s)|.$$

Show that H is compactly embedded in J.

Problem 401 *Suppose that each of H, J, K is a Banach space so that H is compactly embedded in J. Suppose also that $F : H \to K$ is continuous when regarded as a function on J and that $x \in H$, $r > 0$. Suppose in addition that if $\|y - x\|_H \leq r$, and $\epsilon > 0$, there is $h \in H$, $\|h\|_H \leq r$, so that*

$$\lim_{t \to 0+} \frac{1}{t}(F(y + th) - F(y)) = -F(x). \tag{22.2}$$

Show that there is $u \in H$ so that $\|u - x\|_H \leq r$ and $F(u) = 0$.

Problem 402 *Suppose that H, J, K are as in Problem 401, $r > 0$ and*

$$G : \{z \in H : \|z - x\| \leq r\} \to K$$

is continuous when regarded as a function on J. Suppose also that $g \in K$ and that if $\|y - x\|_H \leq r$, then there is $h \in H$ so that $\|h\|_H \leq r$ and

$$\lim_{t \to 0} \frac{1}{t}(G(y + th) - G(y)) = g.$$

Show that there is $u \in H$, $\|u - x\|_H \leq r$ so that

$$G(u) = g.$$

This gives a version of a Nash–Moser inverse function theorem.

Problem 403 *Suppose that F, c are as in Problem 393. Show that the differential equation (22.1) generates a global semigroup if $c = \infty$ for all $g \in X$ and generates a local semigroup if $c < \infty$ for some $g \in X$.*

Problem 404 *Suppose that $\phi : R^2 \to R$ is a C^2 nonnegative-valued function and*

$$F(x) = (\nabla\phi)(x), \ x \in R^2$$

and that

$$(F'(x))^{-1} \ \text{exists for all } x \in R^2.$$

Perhaps think of the graph of ϕ as being a smooth mountain (an idealized version of Stone Mountain in Georgia, Ayers' Rock in Australia or Enchanted Rock in Texas). Consider two strategies for climbing this mountain starting at $(x, \phi(x))$ for some $x \in R^2$:

- $z(0) = x$, $z'(t) = F'(z(t))^{-1}F(x)$, $0 \leq t$.
- $w(0) = x$, $w'(t) = F(w(t))$, $0 \leq t$.

Identify one of these two items as continuous steepest assent and the other as derived from the continuous Newton's method (turned around in direction to seek a maximum). Determine that one method corresponds to the strategy of always climbing in the steepest direction whereas the other has the strategy of always keeping the gradient direction the same as that at the starting point x. Ponder which strategy might be optimal.

Chapter 23
Generalized Semigroups Without Forward Uniqueness

Suppose that p is a nonconstant complex polynomial.

Definition 41 *A trajectory for continuous Newton's method (for the finding roots of p) is a continuous function $z : R \to C$ so that*

$$(p(z))'(t) = -p(z(t)), \ t \in R. \tag{23.1}$$

Denote by Q_p the set of all trajectories for p.

Problem 405 *Suppose that z is a trajectory for p and J is a subinterval of R such that $p'(z(t)) \neq 0$ for all $t \in J$. Show that*

$$z'(t) = -\frac{p(z(t))}{p'(z(t))}, \ t \in J.$$

Problem 406 *Realize that*

$$-\frac{p(w)}{p'(w)}, \ w \in C, \ p'(w) \neq 0$$

is the Newton quotient for p at w and contemplate the nature of (23.1).

Problem 407 *Suppose that $z \in Q_p$, $x \in C$, $z(0) = x$ and z satisfies (23.1). Show that*

$$p(z(t)) = \exp(-t)p(x), \ t \in R.$$

Problem 408 *Show that if $z \in Q_p$, then*

$$u = \lim_{t \to \infty} z(t)$$

exists and $p(u) = 0$.

Problem 409 *Show that every member of C is contained in some member of Q_p.*

J.W. Neuberger, *A Sequence of Problems on Semigroups*,
Problem Books in Mathematics, DOI 10.1007/978-1-4614-0430-9_23,
© Springer Science+Business Media, LLC 2011

Definition 42 *A subset G of C is called an incoming trajectory for p if there is $x \in C, d \in R$ and $z \in Q_p$ so that*

$$p(x) \neq 0, \ p'(x) = 0, \ z(d) = x \text{ and } G = z((-\infty, d]).$$

Definition 43 *Denote by M_p the complement of the union of the set of all incoming trajectories for p.*

Problem 410 *Show that every component (maximal connected subset) of M_p contains just one root of p. Show also that if $z \in Q_p$ and the range of z intersects a component S of M_p, then*

$$u = \lim_{t \to \infty} z(t)$$

is the root of p contained in S.

The following may be of interest in connection with Problem 410:

Problem 411 *Suppose $x \in C$, $p'(x) = 0$ and $p(x) \neq 0$. Show that there is $\delta > 0$ so that if $s \in R$ and $0 < |s| < \delta$, then:*

- *If $s > 0$, there is $f[0, s] \to C$ so that $f(0) = x$ and*

$$(p(f))'(t) = -p(f(t)), \ t \in [0, s].$$

- *If $s < 0$ and $v = -s$, there is $g : [0, v] \to C$ so that*

$$g(v) = x, \ (p(g))'(t) = -p(g(t)), \ t \in [0, v].$$

Problem 412 *Type in and run the following Mathematica routine (here using Mathematica 7, other versions may have a different last line):*

$$p[z_-] := (z^2 + z + 1)(z - 2)$$
$$f[x_-, y_-] := -Re[p[x + Iy]/p'[x + Iy]]$$
$$g[x_-, y_-] := -Im[p[x + Iy]/p'[x + Iy]]$$
$$StreamPlot[\{f(x, y), g(x, y)\}, \{x, -1, 3\}, \{y, -2, 2\}]$$

The first line specifies a polynomial, the last line specifies a plot window (here $[-1, 3] \times [-2, 2]$). These of course can be changed.

Problem 413 *For plots generated by the Mathematica code in Problem 412, take particular notice of members w of C at which $p'(w) = 0$ and $p(w) \neq 0$. Note that there are at least two trajectories entering w and at least two trajectories leaving w. Note that these are points at which forward uniqueness of trajectories is lost. Contemplate what this might mean in view of semigroups.*

Problem 414 *For a given nonconstant complex polynomial p, define T_p with domain $[0, \infty)$ so that if $t \in R$ and $H \subset C$, then*

$$T_p(t)H =$$
$$\{w \in C : \exists z \in Q_p, s \geq 0, \ni z(0) \in H \text{ and } z(s) = w\}.$$

Articulate continuity properties of T_p. Discuss whether T_p should belong to semigroup theory since forward uniqueness is such a venerable part of operator semigroup theory.

Problem 415 *Find and read [44] for some lemmas that may be helpful in solving problems in this chapter.*

Problem 416 *Make a series of plots using the Mathematica code in Problem 412 with the polynomial p replaced by the Riemann Zeta function. Take a variety of computational windows, concentrating on those boxes which contain one or several zeros of the Zeta function or zeros of the derivative of the Zeta function. Note that at zeros of derivatives of the Zeta function, there appear to be two incoming and two outgoing trajectories. I have made a vector field over large windows such as $\{x, -10, 10\}, \{y, -10, 120\}$ and have observed very interesting patterns. It is intriguing that if there were to be a pair of zeros of Zeta positioned symmetrically on either side of the critical line, then this pattern would almost certainly be substantially broken. This suggests a study of such patterns, in the hope that a topological, qualitative approach to the Riemann hypothesis might be uncovered.*

Problem 417 *Try to make a systematic study of semigroups which have trajectories for which there are two or more trajectories which 'leave' a given point along separate paths. Such semigroups seem to have been rarely studied (see [3] for what was my introduction). The general idea of resolvents, Laplace transforms which have figured so substantially in Chapters 5 and 17, for example, has yet to be touched for these semigroups which do not have forward uniqueness.*

Chapter 24
Nonlinear Semigroups and Monotone Operators

Suppose H is a real Hilbert space.

Definition 44 *Suppose A is a function with domain in H and range in the set of all subsets of H. One says that A is monotone if*

$$\langle u - v, x - y \rangle_H \geq 0 \quad \text{if } x, y \in H \text{ and } u \in Ax, \ v \in Ay.$$

Definition 45 *One says that a monotone operator A is maximal monotone if given a monotone operator B, such that*

$$D(A) \subset D(B) \quad \text{and} \quad Ax \subset Bx, \ x \in D(A),$$

then $B = A$. Note that B may be a set-valued transformation.

Problem 418 *For the semigroup T of Problem 12 define*

$$A \ = \ \{(x,y) \in [0,1] \times R : y = \lim_{t \to 0+} \frac{1}{t}(T(t)x - x)\}.$$

Show that A is the negative of a monotone transformation.

Problem 419 *For A as in Problem 418, find a maximal monotone operator B with domain $D(A) = [0,1]$ which contains $-A$ in the sense that*

$$\text{if } x \subset [0,1], \text{ then } \ -Ax \in Bx.$$

Problem 420 *Suppose A is a maximal monotone operator with domain in H. Show that if $\lambda > 0$, then*

$$\text{the range of } I + \lambda A \text{ is all of } H$$

in the sense that

$$\cup_{x \in D(A)}(I + \lambda A)x \ = \ H.$$

J.W. Neuberger, *A Sequence of Problems on Semigroups*,
Problem Books in Mathematics, DOI 10.1007/978-1-4614-0430-9_24,
© Springer Science+Business Media, LLC 2011

Problem 421 *With A as in Problem 420, show that*

$$(I + \lambda A)^{-1}, \lambda > 0,$$

exist as ordinary functions with domain all of H.

Problem 422 *Suppose A is a monotone operator with domain in H. Show that there is a maximal operator B such that*

$$D(B) = D(A)$$

and that if $x \in D(A)$, then

$$A(x) \subset B(x).$$

For Problems 423–426, suppose that T is a strongly continuous nonexpansive semigroup on a closed convex subset Ω of a Hilbert space H. One might consult [9], pages 114–120, for some additional lemmas for these problems.

Definition 46 *If $t > 0$, denote*

$$\frac{1}{t}(I - T(\lambda))$$

by A_t.

Problem 423 *Show that if each of $\lambda, t > 0$, then*

$$(I + \frac{\lambda}{t} A_t)^{-1}$$

exists and is nonexpansive.

Problem 424 *Show that if $\lambda > 0$, then*

$$\lim_{t \to 0+} (I + \frac{\lambda}{t} A_t)^{-1} x$$

exists for all $x \in \Omega$.

Problem 425 *Suppose that A is a maximal monotone operator on Ω and $x \in D(A)$. Show that there is an element in Ax of minimum norm.*

Definition 47 *Suppose A is as in Problem 425. Denote by A_0 the transformation whose domain is the same as the domain of A, but for $x \in D(A)$, $A_0 x$ is the element of the set Ax with minimum norm.*

Problem 426 *Show that there is a maximal monotone operator A on Ω so that if $x \in D(A)$, then*

$$\lim_{t \to 0+} \frac{1}{t}(I - T(t))x = A_0 x.$$

Definition 48 *Suppose now that A is a maximal monotone operator. Denote*

$$(I + \lambda A)^{-1} \text{ by } I_\lambda.$$

Problem 427 *Using the notation of Problem 422, show that if $\alpha \geq \beta > 0$, then*

$$I_\alpha x = I_\beta(\frac{\beta}{\alpha}x + (1 - \frac{\beta}{\alpha})I_\alpha x), \ x \in H.$$

Problem 428 *Suppose A is as in Problems 422, 427, and $\alpha \geq \beta > 0$. Define*

$$\lambda = \frac{\beta}{\alpha} \text{ and } \mu = 1 - \frac{\beta}{\alpha}$$

and

$$\phi_{k,j} = \|I_\beta^j x - I_\alpha^k x\|, \ k, j = 1, 2, \ldots.$$

Show that

$$\phi_{k,j} \leq \lambda \phi_{k-1,j-1} + \mu \phi_{k,j-1}, \ k, j = 1, 2, \ldots.$$

In preparation for Problem 433 we have an exercise in which one may use various ideas in probability which are useful in Problem 434. Also we see another example in which we can apply ideas from probability theory in the development of semigroup theory. We saw in Chapter 21 an example in which one can apply semigroup theory to probability theory.

Problem 429 *Solve numerically the problem of finding u with domain the subset of*

$$\Omega = [0, 1] \times [0, 1]$$

such that

$$u_1(x, y) = u_2(x, y), \ (x, y) \in \Omega, \ u(0, y) = f(y), \ y \in [0, 1], \qquad (24.1)$$

where u_1, u_2 denote, respectively, partial derivatives with respect to the first and second arguments of u and f is a C^1 function with domain $[0, 1]$. Choose an integer $n > 2$ and divide the interval from $(0, 0)$ to $(0, 1)$ in n equal subintervals. Choose x such that $0 < x \leq 1$ and define

$$v_{0,j} = f(j/n), \ j = 0, 1, \ldots, n.$$

With $\delta = 1/n$ and $h = x/n$ define inductively

$$v_{i,j}, \ j = 0, \ldots, n - i, \ i = 1, 2, \ldots, n$$

such that

$$\frac{v_{i,j} - v_{i-1,j}}{h} = \frac{v_{i-1,j+1} - v_{i-1,j}}{\delta}, i = 1, 2, \ldots, n, j = 0, \ldots, n - i.$$

Show that

$$v_{n,n} = (B_n^f)(x) = \Sigma_{k=0}^n \binom{n}{k} x^k (1-x)^{n-k} f(k/n).$$

Problem 430 *Show that in the setting of Problem 429, if $x \in [0,1]$, then*

$$\lim_{n\to\infty} B_n^f(x) = f(x).$$

It is convenient to know about central limit theorems in order to solve this problem, but not necessary.

Problem 431 *Show that if u satisfies (24.1), then*

$$u(x,y) = f(y+x), \; x,y \in [0,1], \; x+y \leq 1$$

which is want one wants if this finite difference method works for Problem 429.

Problem 432 *Show that the convergence indicated in Problem 429 is uniform on $[0,1]$. The polynomials B_n^f above are the famous Bernstein polynomials. The result in the present problem gives Bernstein's proof of the Weierstrass Approximation Theorem, which says that any continuous function on an interval is the uniform limit of a sequence of polynomials.*

Problem 433 *Using the notation of Problems 422–428, show that if $t \geq 0$ and $x \in \overline{D(A)}$, then*

$$\lim_{n\to\infty} (I_{t/n})^n x \quad exists. \tag{24.2}$$

Problem 434 *Suppose S is a function with domain $[0,\infty)$ and range the set of all functions from X to X defined by*

$$S(t)x = \lim_{n\to\infty} (I_{t/n})^n x, \; t \geq 0, \; x \in \overline{D(A)}.$$

Show that S is a strongly continuous semigroup of contractions over $\overline{D(A)}$.

Problem 435 *Show that the semigroup S in Problem 434 may be uniquely extended to all of X by continuity and that the resulting extension is a strongly continuous contraction semigroup on X.*

This is the main part of the famous theorem of Crandall–Liggett [12]. See the notes in Chapter 25 for references, historical information and additional problems.

Problem 436 *Find and read [9],[60] for more background on problems in this chapter.*

Chapter 25
Notes

Some General Comments

Semigroups, semiflows, semidynamical systems—all are the same. Once when I gave a talk at the University of Alaska in Fairbanks, I mentioned that I once heard that the Inuit had something like eighty different words for snow. One of my hosts countered that the Inuit had over a hundred words for snow. So much for a Texan making comments about snow to Alaskans! Anyway, there being three words—semigroups, semiflows, semidynamical systems— for the same thing is indicative of the importance of a certain idea—that of a one-parameter family of transformations T so that for some space X,

$$T(0) = I, \text{ the identity transformation on } X, \ T(t)T(s) = T(t + s), \ t, s \geq 0$$

$$(25.1)$$

and

$$T(t) : X \to X, \ t \geq 0.$$

Now the origins of these three terms come from different mathematical cultures. The term 'semigroup', as used in this book, arose from abstraction of time-dependent autonomous partial differential equations. 'Semiflow' seems to me to have arisen from topology whereas 'semidynamical system' has a life of its own in the vast world of dynamical systems. The terms 'semiflow' and 'semidynamical system' are used infrequently, for objects in this list of problems, compared to the term 'semigroup'. For each of the three terms, dropping the 'semi' indicates that the members $T(t)$, $t > 0$ all have inverses, commonly continuous ones, and that the semigroup law (25.1) extends to all of R. 'Semigroup', on the other hand, is more pervasive than 'group' in the context of one-parameter families of transformations. A reason for this is that time-dependent partial differential equations commonly are not reversible in time. Physically, knowing the heat distribution in a metal bar doesn't tell us much about how hot the bar was an hour ago. Actually, for T the heat equation semigroup $T(t)$ *is* invertible for $t > 0$, but although $T(t)^{-1}$ exists it is only densely defined and is discontinuous at each point at which it is defined. For plenty of examples of strongly continuous semigroups T, $T(t)$ is

J.W. Neuberger, *A Sequence of Problems on Semigroups*,
Problem Books in Mathematics, DOI 10.1007/978-1-4614-0430-9_25,
© Springer Science+Business Media, LLC 2011

not invertible for some $t > 0$; for example the semigroup T on X, where X is the Banach space of bounded continuous functions on $[0, \infty)$ with sup norm and

$$(T(t)(f))(x) = f(t + x), \ t, x \geq 0.$$

Many things in life are not reversible, at least according to my own experience. Confining oneself to time-dependent partial differential equations which are reversible means one misses out on a lot. This said, however, there are many important time-dependent processes which *are* reversible, a most noteworthy one being the Schrödinger equation of quantum mechanics.

Something that conventionally separates things usually called 'semigroup, semiflow' or 'semidynamical system' is that generators of 'semigroups' are commonly densely defined everywhere discontinuous transformations. Such generators are not commonly found in papers dealing with semidynamical systems or semiflows, in my experience.

The recent incorporation of 'local semigroups' into existing semigroup theory (see Chapter 17) seems to be a significant extension but it is hardly new in that Sophus Lie considered them in his quest for integrating factors for systems of ordinary differential equations. My own introduction to Lie's work came in a graduate seminar that I was, many years ago, conducting at Emory University. We were trying to extend to non-locally compact spaces some of von Neumann's work, the part which was seminal to the eventual solution of Hilbert's Fifth Problem. I mentioned that one of us should find out what Sophus Lie did. It turns out that no one took me up on this, so I made the attempt myself, studying essentially work from [27]. Some five years later, in about 1970, I made use of my inspiration from Lie's work in [36], but it wasn't until my collaboration with Dorroh that the idea came to fruition as an alternative to the (stalled, in my opinion) theory of nonlinear semigroups (as a generalization of existing linear semigroup theory).

The present volume of these notes has its origins, [46], in notes I wrote in Spanish for the XIII Escuela Venezolana Matemàticas at the Universidad de los Andes in Merida, Venezuela, 6–15 September 2000. These notes comprised about 48 pages and contained about 111 problems. In the course the students (faculty, graduate students and a few undergraduates) vigorously attacked and settled many of the problems in these notes, leading to many intense and enlightening discussions. The present form of these notes started with my translation of the original notes into English. The number of problems is nearly quadruple that of the original notes. There are several chapters of problems dealing with subjects which didn't even exist in 2000.

The original notes benefited greatly, for both mathematics and Spanish, from the help of Alfonso and Miryam Castro, Mario Jimenez, Barbara Neuberger, Víctor Padrón, María Mera Rivas and María Cristina Trevisán.

I particularly thank the organizing hosts, Victor Padrón and Oswaldo Arajo, for their great hospitality and their great effort in arranging for this

Summer School. It was a long-cherished dream to be able to conduct a course using only Spanish, both in class and outside the classroom. This course represented a particular challenge since I was not lecturing but rather constantly interacting with the participants while they were presenting arguments. Explaining a flaw in someone's argument is a challenge even in one's native language. Everyone was very generous in putting up with my limited Spanish.

For some problems in semigroups, complex analysis enters in an essential way. Chapter 23 is one instance. The result of Beurling in Chapter 21 is another instance. The subject of holomorphic semigroups is generally complex-based. Earlier work (see in particular [21], for example) is based on Laplace transforms as in Chapter 17. Passage from semigroup to resolvent is frequently presented as a matter of inverting a Laplace transform—something often done using contour integrals but this is not done in the present volume.

Problem 437 *Find corresponding complex field results for the (majority) of problems in this book which are stated (usually implicitly) for the real field.*

25.1 Notes on Chapter 2

Problem 2 is the earliest functional equation of which I know. It is remarkable that it has a vast set of solution, if the hypothesis of continuity is omitted but only a simple family of solutions if continuity is assumed. Continuity, it turns out, implies differentiability. Problem 5 urges a reading of at least the second half of Hilbert's Fifth Problem. It was Hilbert who likely was the first to understand the profound power of combining algebraic and topological hypotheses in the presence of the possibility of analytic results. Almost all of the problems in this volume owe a debt to this legacy.

Problems 10, 12 are two examples considered at the start of the quest to generalize linear semigroups to nonlinear semigroups. These early examples led to [35] and attempts to incorporate the idea of resolvent into nonlinear study. See [9], [60], [29] and references contained therein as well as notes in Section 25.23 for more details.

The terms 'continuous' and 'strongly continuous' are somewhat misleading, but completely standard, when applied to semigroups. The term 'continuous' is a stronger notion than is 'strongly continuous'. To add to confusion, there is also 'weak continuity' of semigroups which refers to continuity with respect to a weak topology.

25.2 Notes on Chapter 3

Operator semigroup theory has a curious property that often results from a special case are applicable to more general cases. Many of the ideas developed in this chapter for translation semigroups have direct application to much more general cases. This holds true especially in Chapter 17 in which linear theory is *applied* to nonlinear theory.

For some decades, a thrust was to try to develop a nonlinear theory in *analogy* with linear theory. This led to many interesting developments but to this day has had a rather limited success. Generalized translation semigroups (see Chapter 17) ultimately gave a fairly satisfactory theory. For this reason alone, translation semigroups would be of considerable interest. Nonlinear semigroups in Chapter 17 give rise to linear semigroups which are essentially translation semigroups on a metric space.

In this chapter, some probability distributions arise in a natural way. Someone working their way through these problems has at least two choices. One is to find a source of information on Poisson distributions. Problem 36 is then essentially a consequence of an appropriate central limit theorem. The other choice is to closely study the distributions indicated in Problem 34 to see directly that as $n \to \infty$, then the sequence of distributions, for some $\lambda > 0$, converges to a stepfunction which is zero from $[0, \lambda)$ is $\frac{1}{2}$ at λ and is one on (λ, ∞). The same distribution appears in an essential way in Chapter 17, so effort spent on the Poisson distribution here will be rewarded later.

My own introduction to the application of probability to semigroup theory stems from my encountering Bernstein polynomials in what is outlined in this book as Problem 429. In 1958 while teaching my first graduate course. I rather idly was looking into how numerics worked out for certain simple partial differential equations. Much to my surprise, the Bernstein polynomials suddenly arose. I knew little of central limit theorems then and, before that time, Bernstein polynomials looked strange to me. My brute force approach in showing convergence of the numerical scheme in Problem 429 led me in a life-long affinity for how probability, semigroups and partial differential equations relate. This episode also led me to a study of quasianalyticity in terms of higher order differences, as indicated in Chapter 21. It is usually hard to put in a good word for ignorance, but in this case my lack of knowledge of central limit theorems led me to some nice things.

25.3 Notes on Chapter 4

Continuous semigroups are very special cases of semigroups of linear transformations. They are essentially based on ordinary differential equations in a Banach space. Continuous semigroups are essentially infinite-dimensional generalizations of constant coefficient systems of linear equations, but many of the problems in this chapters reveal things which help in the study of more general semigroups—those which do pertain to partial differential equations.

One result of working the problems in this chapter is to see in a rather direct way how differentiability may arise from algebraic semigroup properties taken together with continuity. The identity in Problem 38 is very useful in this regard. Problem 42 is an early chance for a reader to see a generator appearing for a semigroup.

A possible strategy in gaining understanding of the matters of this chapter is to follow through using the semigroup g from Problem 3.

Problem 46 indicates how Picard's method of successive approximations leads to existence of a semigroup with generator $B \in L(X, X)$, X a Banach space. Remaining problems in this chapter show how exponentials of such a transformation B represent the semigroup with generator B. Problems 49, 50 give a product expression and the equivalent series expression, respectively.

25.4 Notes on Chapter 5

As already indicated, linear semigroups which are strongly continuous but not continuous form the theoretical basis for autonomous time-dependent linear partial differential equations. This subject had its start with the work of Marshall Stone on the Schrödinger equation in the 1930s, showing that this equation gave rise to a group. The massive work of Hille and Phillips [21] in the 1950s developed the theory of strongly continuous linear semigroups into something close to its present form. The book [21] is an excellent reference, but one might still seek out Hille's original [20] book of the same title which has more concrete information about partial differential equations. A prized possession of mine is a copy of Hille's *Functional Analysis and Semigroups* given to me by Phillips.

The books [17], [19], [16] and [59] deal with linear semigroups and have a good deal of information on applications. In [66] there is an excellent chapter on strongly continuous linear semigroups.

To anyone reading any of these references it will be clear that problems in the present book contain just an introduction to the study of one-parameter semigroups.

25.5 Notes on Chapter 6

The heat equation gives the premier example of a semigroup that comes from a time-dependent PDE. In a sense it dates back to Fourier. One can solve the heat equations by Fourier's method, 'separation of variables', and then compare results with numerical solutions. Of particular interest here is how the relationship between 'implicit' and 'explicit' methods for solving the heat equation has its counterpart in the general theory of linear strongly continuous semigroups in that two plausible exponential formulae have strikingly different levels of viability (see Problem 84). Contemplate the First Law of Numerical Analysis, Chapter 6, in this regard.

25.6 Notes on Chapter 7

Definition 8 is for Fréchet derivative. Later problems give some properties. Problem 97 gives a local existence and uniqueness theorem for ordinary differential equations. It can be proved by the method of successive approximations, as in Problem 46. Problem 98 gives a limit theorem that is needed in Chapter 14. Problems 100, 102 give two versions of the spectral theorem that are useful in connection with Chapter 16. Problems 100, 101 give a generalization of the fact that in finite-dimensional Euclidean space, a symmetric linear transformation has a basis in terms of which the transformation has a diagonal matrix representation. A reference such as [66] might be consulted for lemmas and background.

The present chapter also contains some preliminaries to problems in Chapter 16 in which a single linear transformation T has two (related) adjoints. In this case, one adjoint of T is continuous and the other adjoint is not. This is characteristic of Sobolev gradients arising from problems in differential equations. It is helpful for a reader to reconcile this pair of adjoints. A reader might see later how gradients essentially based on one definition of adjoint lead to viable numerical methods whereas when based on ordinary gradients become a disaster (see Chapter 14 and [43] for problems on this issue).

For more background on Problems 100, 102, see, for example, [66] or other books that deal with spectral theory.

25.7 Notes on Chapter 8

Results on combining two (or more) continuous linear semigroups give some indication of what to expect for combining two strongly continuous linear semigroups and also for combining two nonlinear semigroups. A starting point is to note that for $n > 1$, a positive integer, and $A, B \in L(R^n, R^n)$, then it does not necessarily follow that

$$e^{tA}e^{tB} = e^{t(A+B)}, \ t \geq 0,$$

but some aspects of this law of exponents can be regained by

$$e^{t(A+B)} = \lim_{k \to \infty} (e^{\frac{t}{k}A}e^{\frac{t}{k}B})^k.$$

25.8 Notes on Chapter 9

In regard to Problem 144, I seem to recall working this out sometime around the 1970s, but have been unable to recover my argument. A path to my original scratchings on this problem might be found by first doing a web search on 'A Guide to the J. W. Neuberger Papers'. My actual papers up to 2002 (more to be added later) are in the

Dolph Briscoe Center for American History

at the University of Texas.

It might be easier to just work this out for yourself than to find it. I would not be surprised to find that others have found this result, but I cannot offer a reference.

What might be new to some is the introduction of a bit of near-ring theory, which seems rarely used in analysis problems.

The site indicated above contains many of my scratchings from the early 1960s to about 2001. I don't particularly wish it upon someone to spend a lot of time searching these papers, but they are there if for some reason someone might want to attempt to reconstruct some of my (usually vague at best) mathematical thoughts.

In [62] there are results by Coke Reed on combining dynamical systems. Two continuous dynamical systems T and S on a Banach space X are said to combine provided that if $x \in X$ and $a, b \in R$, then there is $y \in X$ so that if $\epsilon > 0$ there is $\delta > 0$ so that if $t_0, t_1, \ldots, t_n$ is a partition from a to b of mesh less than δ, then

$$\| \left(\Pi_{k=1}^n \left(T \left(t_k - t_{k-1} \right) S \left(t_k - t_{k-1} \right) \right) \right) (x) - y \| < \epsilon. \tag{25.2}$$

If there is a continuous dynamical system U on X so that

$$K(b - a)x = y,$$

y as in (25.2), then it may be said that T, S combine to get U. The paper [62] contains a number of results on combining dynamical systems and surprisingly, contains counterexamples to some combination conjectures which struck me as plausible and probably true, but are not true.

Problem 438 *Obtain and study [62]. Ponder how results there show some limitations on combining semigroups as well as some promising directions of inquiry. Examine other papers of Coke Reed as found in MathSciNet.*

The paper [62] and others by Coke Reed show connections between the semigroup-dynamical system-flow cultures. (Disclaimer: I introduced Coke Reed to the study of dynamical systems in the 1960s. A reader might find it interesting to learn where his study of dynamical systems has led him.)

25.9 Notes on Chapter 10

This chapter contains a number of results which are an extension of Chapter 5 and are preliminary to Chapter 11. Perhaps the fundamental difficulty in dealing with strongly continuous linear semigroups that are not continuous, is that the generators of such semigroups are always closed densely defined linear transformations. Such transformations are always discontinuous at each point at which they are defined, but many calculations which are natural for continuous linear semigroups can be arrived at by rather convoluted reasoning (if they are true at all). Extensive use of resolvents of generators of strongly continuous linear semigroups is characteristic of many of the main developments in the theory. The present chapter gives some developments which will be used in Chapter 11. What is probably new to even seasoned researchers in the field is the use of probability measures as in Problem 155, again the Poisson distribution as introduced in Chapter 3.

25.10 Notes on Chapter 11

Results of the present chapter give an outline of some special cases of the celebrated Trotter–Kato development, which is of great interest in partial differential equations. Developments in this chapter follow the outline given in Chapter III, Section 5 of [17], but there are substantial differences in the present development, particularly in the use of probability distributions

$$\phi_{m,\lambda}, \ m \in Z^+, \ \lambda > 0. \tag{25.3}$$

So far as I know, these distributions were first applied to semigroup theory in [14]. In the present volume, these distributions are used in Chapters 5 and 3, but their use there was taken from [14].

Feynman–Kac formulae, based on Trotter–Kato developments, are of interest in quantum mechanics. There is an account of this in [19] in which there is indicated a continuing mystery concerning these formulae. In Chapter 12 there are some problems involving applications of Trotter–Kato formulae to the numerical solution of time-dependent partial differential equations.

I consider Problems 179 to 181 rather speculative. My main reason for including these is the following: Consideration of results surrounding (25.3) might lead to a new and more comprehensive family of arguments for semigroups, those more based on ideas from probability, than have been usual in semigroup theory. Traditional Trotter–Kato results seem to be based on arguments such as those found in Chapter III, Section 5 of [17]. It just might be that some of Problems 179 to 181 can be proved true using these newer applications of probability to semigroup theory and consequently extend the theory.

In my opinion, to show that two semigroups T, S combine to give a third semigroup U is more important than the question of how possible generators of T, S can be used to gain a generator of U even though the second issue is certainly a significant one.

25.11 Notes on Chapter 12

Splitting methods are widely used in practice, often beyond the realm of established proofs indicating their validity. There is a considerable literature that may be consulted. The splitting method is a good way to deal with reaction–convection–diffusion equations. Each of the reaction, convection and diffusion equations has their own highly efficient method of solution. One main time step in the splitting method for such equations entails a sequence of three substeps: the first using only a diffusion method, the second a convection method and the third a reaction method.

25.12 Notes on Chapter 13

The idea of using a method of steepest descent to find zeros or critical points of real-valued functions goes back at least to Cauchy. This chapter gives some basic results concerning zeros of linear transformations between two Hilbert spaces. The two spaces may be Sobolev spaces, in which case results apply directly to systems of linear differential equations. In appropriate finite-dimensional spaces, results apply to finding numerical approximations of solutions to such equations, as illustrated in Section 25.13. A reference for problems in the present chapter is [43], Chapter 3.

25.13 Notes on Chapter 14

In [43] there is a fairly complete recent discussion of Sobolev gradients. There are applications to problems of transonic flow, minimal surfaces and super-conductivity ([64], [65]). The gradient inequality (Definition 17) is a fairly strong hypothesis and I hope that future work will seek to replace it with weaker conditions.

A main issue relating semigroups and steepest descent is the following: Given a real-valued C^1 function ϕ on a Hilbert space X and a gradient $\nabla\phi$ for ϕ, that is, a function which satisfies

$$\phi'(x)h = \langle h, (\nabla\phi)(x)\rangle_X, \; x, h \in X,$$

determine conditions on ϕ so that if

$$z(0) = x, \; z'(t) = -(\nabla\phi)(z(t)), \; t \geq 0,$$

then

$$\lim_{t\to\infty} u = \lim_{t\to\infty} z(t) \text{ exists}$$

and

$$(\nabla\phi)(u) = 0.$$

There is the broad problem:

Problem 439 *Make an investigation of the gradient inequality and try to find weaker conditions which imply the conclusion of Problem 227.*

One can consult [66] for example, and Chapter 16, to obtain more information about the projections onto

$$\{\begin{pmatrix} x \\ Tx \end{pmatrix} : x \in D(T)\},$$

where H, K are Hilbert spaces and T is a closed densely defined linear transformation on H with range in K. The original formula was due to von Neumann [70]. These projections serve as a point of departure in the construction of Sobolev gradients. See also [1] as a general reference for Sobolev spaces.

This chapter gives only the barest introduction to Sobolev gradients, but [43] and references contained therein give a fairly comprehensive recent account and a bibliography. A reader might google 'Sobolev gradient' or 'Sobolev gradients' for a further impression of this subject.

Continuous steepest descent using Sobolev gradients leads naturally to nonlinear semigroups if the underlying problem is itself nonlinear. A dominant feature is that a properly formulated least squares or variational principle problem leads to a *continuous* nonlinear semigroup, not just a strongly continuous one. Hence the underlying steepest descent equation is essentially an *ordinary* differential equation. The contrast with more conventional formulations is illustrated by minimal surface problems: In [8], for instance, the process for finding a minimal surface is 'evolution by mean curvature' of a conventional time-dependent partial differential equation. A corresponding Sobolev gradient approach yields a steepest descent, with continually varying metric, which is an ordinary differential equation in function space (see [43], Chapters 11 and 16). This fact alone seems to justify interest in Chapter 9 which deals with continuous nonlinear semigroups.

This raises a question of both a numerical approach and a theoretical approach, using Sobolev gradients in both cases, to such problems as the Nash embedding problem and the Poincaré conjecture. These are, at present, descent or conventional Newton's method processes which can be embedded into continuous processes giving rise to effectively time-dependent partial differential equations.

Problem 440 *Try to formulate the two problems in the above paragraph in terms of Sobolev gradient descent in order that the process may be considered as a problem concerning ordinary differential equations in a function space.*

In a sense, such a corresponding pairing between time-dependent partial differential equations and Sobolev gradient ordinary differential equations in function space, is already illustrated in contrasting methods for Moser's inverse function theorem (Problem 402 of Chapter 22).

25.14 Notes on Chapter 15

A great deal of computational effort continues to be expended on numerical solution of time-dependent partial differential equations. Equations of Navier–Stokes, which govern a wide variety of fluid flows, are a notable example.

Semigroups as developed in these notes grew out of, and remain as, an abstraction of autonomous time-dependent partial differential equations. Even for problems, such as elliptic systems, which do not physically involve time, a semigroup is associated. For example, if X and Y are Hilbert spaces and $F : X \to Y$ is such that the problem of finding $u \in X$ such that

$$F(u) = 0 \qquad (25.4)$$

represents a system of partial differential equations, then critical points of ϕ, defined by

$$\phi(x) = \frac{1}{2}\|F(x)\|_Y^2, \ x \in X,$$

can be turned into a semigroup problem by means of

$$z(0) = x, \ z'(t) = -(\nabla \phi)(z(t)), \ t \geq 0. \qquad (25.5)$$

Numerical calculations such as those introduced here are at once a practical matter aiming to get concrete information about solutions, and also a means to gain insight into the theory of semigroups. Literature on numerical solution of time-dependent PDEs is truly vast, running from mathematics to physics, chemistry, various branches of engineering, biology and economics. No attempt is made here to do justice to this immense and important area. When (25.5) arises from (25.4), relevant gradients put the problem in somewhat different terms from conventional time-dependent problems, but semigroup issues remain for such problems.

In a practical sense, the theory of Sobolev gradients gives an organized way to determine and to compute preconditioners to be applied to ordinary gradients. Following the theory, preconditioners respect boundary and other supplementary conditions, even nonlinear conditions. Generally the ratio

$$\frac{\text{number of iterations needed using ordinary gradient}}{\text{number of iterations needed with Sobolev gradient}}$$

goes, for a given problem, to infinity as mesh size approaches zero. This is documented in [69], as well as a number of other papers authored or coauthored by Sultan Sial and in references contained in his work.

The above describes numerical symptoms when using Sobolev gradients as opposed to ordinary gradients in descent process for partial differential equations. Some issues may be illustrated by a simple example: Consider a least squares formulation for the simple problem of finding u on $[0, 1]$ so that

$$u' - u = 0.$$

The least squares formulation is essentially (14.5):

$$\phi(u) = \frac{1}{2} \int_0^1 (u' - u)^2, \tag{25.6}$$

for u in what space? The space $L_2([0, 1])$ is not good since then ϕ would be only densely defined and everywhere discontinuous where it is defined, giving a poor or nonexistent gradient. However, the ordinary gradient for a discrete version of (25.6) is in a sense trying to do just that. If, however, the Sobolev space $H^{1,2}([0, 1])$ is used, then the resulting ϕ is continuous and differentiable—in fact it is a quadratic polynomial.

The fundamental idea of Sobolev gradients is this: For finite-dimensional emulations of least square (or energy functional) emulations, the finite-dimensional gradient should be taken with respect to a norm which emulates the theoretical norm which renders the functional in question at least a C^1 functional. This idea can be viewed as a consequence of the basic law of numerical analysis, given in Chapter 6, essentially saying that sensitivities in a functional should be matched by sensitivities in a gradient with respect to which steepest descent is being taken.

25.15 Notes on Chapter 16

References for this chapter are [25] and [43], Chapter 5, Section 5.4. Most of the material of this chapter comes directly from the first reference, as summarized in the second reference. The main object here is the systematic use of the embedding operator between two abstract Hilbert spaces H and H' where the points of H' form a dense subset of H and the norm in H' dominates the norm in H.

The material in this chapter had its origin in the seminal work [6] (reprinted in [5]), in which a 'kernel free' development of potential theory was presented. There are other interesting connections with semigroups in these references. The material of this chapter is essentially an abstract extension of [6] without specific reference to various measures used in potential theory.

The reference [25] contains a solution to an abstract symmetric version of the Kato conjecture, [2]. It is related to Problem 257 by observing that

$$H_1 = H'.$$

An application of the idea in Problem 261 may be found in [53].

This chapter concludes with problems leading up to a formula of von Neumann which applies the development in the first part of the chapter. This formula presents in a simple form the orthogonal projection onto

$$\begin{pmatrix} x \\ Tx \end{pmatrix}, \ x \in X,$$

where T is a closed densely defined linear transformation of X into Y.

Problem 441 *In* (16.2), *describe why one might write*

$$T^t(I + TT^t)^{-1} \ instead \ of \ (I + T^tT)^{-1}T^t.$$

Both of these expressions are linear and continuous. They agree on a dense subset of X. Are they precisely the same?

25.16 Notes on Chapter 17

Since the middle of 1950 some of us have sought a complete theory of non-linear semigroups which has the power of the theory of strongly continuous linear semigroups. Perhaps the first paper in this direction was [35]. The book [9] has a good description of the case of strongly continuous semigroups of contractions on a convex subset of a Hilbert space. The books [60], [68], [29] deal with various extensions to spaces more general than Hilbert spaces. After 1971 or so there has been little substantial progress in the direction of a complete theory although many interesting results had been found. In this context, 'complete theory' means a theory in which a collection of semigroups SG, a collection of generators GEN, and a means of (a) for all elements of SG finding a member of GEN by means of differentiation at zero and (b) for all members of GEN, constructing by means of an exponential formula of a member of SG. The problems in this chapter, for the most part, are from [13], [14], [15].

In [71], von Neumann and Koopman consider Hamiltonian systems on a region Ω in a complex finite-dimensional space. Such systems are commonly a system of nonlinear ordinary equations. They take, using our present terms, a linear representation on complex $L_2(\Omega)$. This representation, using special features of Hamiltonian systems, turns out to be a strongly continuous group of unitary transformations, T. The generator of T turns out to be iA, where A is an unbounded, densely defined self-adjoint linear transformation on $L_2(\Omega)$. A spectral analysis using the spectral theorem, indicated in Chapter 7, is then related to dynamical properties of the Hamiltonian system. This work of von Neumann and Koopman gave encouragement to Dorroh and myself. M. G. Crandall indicated to me (private communication) that he and A. Pazy considered a study of nonlinear semigroups using linear representations, but indicated that they did not pursue this direction. I also had a private communication from G.-C. Rota that he once considered such a study. So, despite the work of Sophus Lie, [27], in the 1800s, work in [71] and the intense, but short, period of excitement briefly described in Chapter 24, it seems that using linear representation as a cornerstone to study nonlinear semigroups had to wait until [38] to seriously begin. After that paper, published in 1973, I occasionally returned to the subject over the next nearly 20 years, with not much in the way of results to report. It wasn't until [14] and its predecessor paper [13] that this direction of research finally made some significant progress in the early 1990s. I will briefly relate how that got started:

In 1992, there was a meeting on semigroups in Curaçao, organized by Jerry Goldstein. It was a small meeting with no parallel sessions. One afternoon there was scheduled a problem session. Several people presented some problems, but then there was silence. I volunteered to present work related to [38]. Bob Dorroh started asking some penetrating questions: 'Is my space X assumed to be locally compact?' for one (my answer was 'no'). This started a period of intense collaboration which eventually resulted in [14]. He knew some relevant things that I would have never 'Dreampt of in my philosophy' (adapted from William Shakespeare's Hamlet, Act I, Scene V). Work of Dennis Sentilles [67] was of crucial use for us. Sentilles was a Ph.D. student of Dorroh in the early 1970s; Dorroh and I have known each other since the 1960s. We could have had our conversations 20 years earlier!

Suppose that T is a jointly continuous semigroup on a subset X of a Banach space. Suppose also that T has a conventional generator B and also Lie generator A. A possible relationship between A and B is

$$(Af)(x) = \lim_{t \to 0+} \frac{f(T(t)x) - f(x)}{t}) = f'(x)Bx, \ x \in D(A), \qquad (25.7)$$

assuming sufficient differentiability is available (otherwise the right side of (25.7) should be written as the directional derivative of f in the direction Bx, all evaluated at x). It is becoming clearer that zeros of B are important in an analysis of A, which in turn is crucial for an analysis of T from our present point of view. This will be made a little clearer in Chapter 19. However, using the material of Chapters 17, 18, 19 is still, for concrete applications, very much in its preliminary stages. I expect a good bit of progress will be led by numerical experiments, which in turn are in their very preliminary states.

In [57] G. E. Parker gives a way to recover semigroups from certian inverse limit sets. This initiated a still largely unexplored alternative way to associate a kind of generator with a nonlinear semigroup. This development merits additional attention. See also Parker's work in [56], [57], [58].

Problem 442 *Find and read [57], [56] to encounter a unique view of linear and nonlinear semigroup theory.*

Additional historical comments relevant to this chapter are in the Notes to Chapter 24.

25.17 Notes on Chapter 18

In a sense, for a nonlinear semigroup T on a Polish space X, the associated semigroup U on the indicated space of measures is more closely related to T than is the representation S of Chapter 17. These semigroups of measures remain, as of this writing, almost entirely unexplored.

If $T, X, B(X), MCR(X)$ are as in Chapter 18, define

$$(U(t)\mu)(\Omega) \; = \; \mu(T(t)^{-1}\Omega), \; t \geq 0, \Omega \in B(X),$$

where $\mu(T(t)^{-1}\Omega) = 0$ if there is not $y \in X$ such that $T(t)x \in \Omega$.

Note that if $x \in X$ and δ_x is the Dirac measure centered at x, then

$$(U(t)\delta_x) = \delta_{T(t)x}, \; t \geq 0.$$

So, U restricted to the Dirac measures in $MCR(X)$ is essentially a copy of T itself. Since U is a linear semigroup, I have called U a linear extension of T.

The semigroups T, S, U of Problems 311, 312, 313 pose some questions of considerable importance. Webb's example (Problem 14), in my opinion, put a stop for some time to the search for a complete theory of nonlinear strongly continuous semigroups in terms of conventional generators. The conventional generator for Webb's example (see Chapter 24) had long seemed too sparsely defined to be useful in a generator-resolvent theory for nonlinear semigroups. The material of Chapters 17, 18 gave rise to possibilities of renewing the quest which was stopped for so long by Webb's example.

Problems 311, 312, 313 indicate a careful study of Webb's example in terms of these later developments. So far as I know, as of this writing, no one has attempted to use the suggestions to better understand Webb's example as a step toward a major advance in nonlinear semigroup theory.

Very few extensions or applications have been made of material from this chapter. I suspect that there are many interesting discoveries to be made in this regard.

25.18 Notes on Chapter 19

Problem 443 *Make a theory of local jointly continuous semigroups, on a complete separable metric space X, which is in analogy to the theory of jointly continuous semigroups in Chapter 17.*

A solution of this problem will make a good publication.

This problem appeared in 2000 notes (in Spanish) which were written for the XIII Escuela Venezolana Mathematicas, [46]. Since then, a substantial part of this problem has been solved by the developments described in Chapter 19 (see also [54]). However, a complete characterization of generators A for local semigroups is still lacking.

One of the Clay Millennium prizes, [18], is for establishing which of global and local existence holds for a Navier–Stokes equation in three dimensions. Developments in Chapter 19 yield a possible attack on this problem, since the *form* of a Lie generator A is clear from the form of Navier–Stokes (see (25.7)). In principle, one has only to decide whether there is a positive eigenvalue of the relevant generator A.

The 'dream' of a numerical attack on this problem is already the subject of serious work. The principle of a numerical attack seems clear, but it seems to be a very large computational problem. Even for a local or global nonlinear semigroup, a suitable discretization of the space of continuous functions whose domain is the Sobolev space $(H^{1,2}(R^2))^2$ needs to be made. Even for a rather rough discretization, the dimension of a corresponding problem is large. For a system of three ordinary differential equations the space that needs a discrete approximation is $(H^{1,2}(R^2))^3$, and so on. For a partial differential equation in time and one space dimension a suitable discretization might require a hundred ordinary differential equations in a 'method of lines' approximation. This would entail a numerical version of $(H^{1,2}(R^2))^{100}$. For Navier–Stokes in three space dimensions and time, the dimension of an appropriate approximation space would be vastly larger.

Problem 444 *Estimate the dimensionality of a reasonably close approximating space for time-dependent Navier–Stokes in three space dimensions, using the approach of the present chapter. Assess whether any present-day computer is up to the job of gaining meaningful evidence on the local–global time existence of time-dependent Navier–Stokes in three space dimensions. How long might we have to wait for an adequate computer?*

25.19 Notes on Chapter 20

For a system of partial differential equations, what condition on a solution is necessary and sufficient in order that there be one and only one function satisfying both the condition *and* the system. For many more-or-less standard systems, such additional conditions are well understood, but the more general question remains one of the outstanding unsettled questions in mathematics (I suspect that the question is considered so outrageous that no one has the nerve to give it a name). For some centuries, attention has been focused mainly, but not entirely, on specifying conditions on the boundary of a region on which a system is to be solved. I suspect that this frame of mind has arisen since so many systems arise as the Euler–Lagrange equations of an energy functional. For a given system on a bounded region to arise from an energy functional on some region, it is almost universal that one starts with a supposed critical point of the functional and then after an integration by parts, arrives at the fact that the Euler–Lagrange equations must be satisfied. When one integrates by parts, one is left with an integral around the boundary of the relevant region. What is to be done with these inevitable integrals around the boundary of the region? It is common that conditions are imposed on potential solutions so that functions satisfying these conditions are such that these boundary integrals become zero. Study of boundary conditions for known types of partial differential equations has led to many interesting results, of course, but a fixation on 'boundary conditions', to the exclusion of consideration of other supplementary conditions, seems not to be so productive. Examples of cases in which 'boundary conditions' may not be an appropriate way to pick out unique solutions include transonic flow problems in which nonlinearities determine type. There is usually no way to merely look at such an equation to pick out subregions of ellipticity or hyperbolicity, for example. The equation often has to be solved first in order to determine such regions. If one has only a method which requires boundary conditions to be known before a solution can be attempted, then one is often caught in an unhelpful circular path.

Now from Chapter 14, the method of Sobolev gradients gives a way to find critical points of some 'energy-like' functionals without having to deal with Euler–Lagrange equations. Using gradients derived from Sobolev metrics, both the start of a numerical theory and a theoretical one are indicated. Some reflection yields that this development gives a clue as to how systems of partial differential equations may be dealt with without first determining 'boundary conditions' appropriate to the system.

The main problem remains as to how one might classify the set of all solutions to a given system. The present chapter is a start in this direction. The main hypothesis is that continuous steepest, starting at any point of the underlying space, descent converges to a unique element. Two functions are called *equivalent* provided that when both are used as starting values for continuous steepest descent, then both descents converge to the same solution. In terms of an appropriate Lie generator, from Chapter 17, the relevant equivalence classes are characterized.

Problems in this chapter should be thought of as leading to a point of view on attacking the unsettled problem to which mention is given above. Finishing the project any time soon is unlikely of course, but interesting results can be expected from anyone seriously considering the problem. What is needed now are more examples.

Note the essential use of semigroup theory in this chapter. Note also that the development is essentially a linear one, even though the focus is on systems of nonlinear partial differential equations.

25.20 Notes on Chapter 21

Problem 368 deals with semigroups for denumerable Markov processes. David Kendall [26] was interested in the following question: If one knows one of the transition probability functions $p_{i,j}$ on some bounded subinterval $[a, b]$ of $[0, \infty)$, can one determine what that transition function is over all of $[0, \infty)$? If such a function were to be real analytic, then analytic continuation would determine it everywhere. However, such analyticity has not been established. About the best that is known in the general case is that these transitions functions are C^1. What Kendall, Kato, Beurling, this writer and others discovered is that the behavior of

$$|P(t) - I| \text{ as } t \to 0+,$$

has a striking effect on smoothness of these transition functions. Problem 366 gives that if

$$\limsup_{t \to 0+} |P(t) - I| < 2,$$

then all transition functions for P are analytic away from zero.

The gist of Problem 367 is effectively that if

$$\liminf_{t \to 0+} |P(t) - I| < 2, \tag{25.8}$$

then all transition functions for P lie in some quasianalytic collection. In [41] there is given a more-or-less constructive method, in case (25.8) is satisfied, of determining all of a trajectory from its values on any small interval, thus giving at least a partial solution of Kendall's problem. See [26], [40], [37] for a more in-depth discussion, references and history.

Problem 362 is Beurling's analyticity result which is used in Problem 366. See [4], [7] for an argument if you haven't yet figured out one for yourself. See [40] for some history of Beurling's result. It was this then-unpublished result which motivated me to seek the creation and publication of [4] and [5].

Problem 382 is from [11], which is concerned with analyticity 'away from zero' of strongly continuous linear semigroups. What is stated in Problem 382 is a weaker conclusion than is actually given in that reference. Results in this paper are a substantial generalization of Beurling's results in [7] (and in [5], in a paper of the same title as [7]). I feel that a great deal more can be learned about these 'analyticity away from zero' semigroups.

So far as I know, no progress has ever been made on Problem 392. In Problem 12 there is a nonlinear semigroup on $X = [0, 1]$ where all trajectories other than the zero trajectory start out at a positive number, then they hit and stay at zero. Even linear examples can have such a property:

Problem 445 *Find such a linear example.*

But many nonlinear semigroups have the property that the set of their trajectories themselves form a quasianalytic collection, hence Problem 392.

See [22] for another characterization of analyticity for a strongly continuous linear semigroup.

The first time I talked to David Kendall was in a transatlantic telephone call, a rarity for me in those days, in 1968. He had some questions about my paper [33], a complicated paper complete with some misprints that made it even harder. He had accepted an invitation to speak at a University of London analysis seminar. The only requirement on topic was that the 'details be sufficiently horrible'—no soft analysis here. He had chosen to present [33] for the occasion! I did get to spend the next summer visiting him in Cambridge. Paper [42], many years later, is a big improvement on [33], but it is quite involved too, sorry to say.

As noted in Chapter 21, Problem 363 is the hardest problem I have ever solved. Only Kendall and perhaps just a few others are known to me to have solved it (or its less general but even more complicated earlier version [33]). The result could use more sunlight shed upon it.

25.21 Notes on Chapter 22

The continuous Newton's method is a continuous generalization of the common Newton's method. The phenomenon that if

$$z(0) = x, \ z'(t) = -F'(z(t))^{-1}F(z(t)), \ t \geq 0,$$

then

$$F(z(t)) = \exp(-t)F(x), t \geq 0, \tag{25.9}$$

has no counterpart in discrete Newton's method. The chaotic domains of attraction, for roots of polynomials (Chapter 23), that come with the discrete Newton's method, do not appear in the continuous case, but rather are artifacts of the discretization. Note that a discretization of the continuous

Newton's method yields the ordinary Newton's method when the discretization parameter is one—the damped Newton's method when the discretization
parameter falls in $(0, 1)$.

I once saw a sign on someone's door at Oak Ridge National Laboratory:

'One Man's Error is another Man's Data'

In a way, this sign could be a parody of some step-by-step numerical
calculations.

Problem 398 is a continuous form of a Nash–Moser inverse function theorem, [30]. See [43], Chapter 8, as well as [48], [49], [51] for more results
related to Problem 398. Moser's epic result uses a scale of Banach spaces and
smoothing operators. It is a triumph of intricate analysis. Results related to
Problem 398 give similar results in a vastly simpler fashion. The key to this
simplicity is that in a continuous version of Nash–Moser type results there is
generally no 'loss of smoothness'. I will try to make this clearer:

Suppose F is a function from a Sobolev space H to a Sobolev space K.
Assume that the problem of finding $u \in H$ such that

$$F(u) = 0 \tag{25.10}$$

represents a system of nonlinear partial differential equations, say of order m.
A step, starting with $u \in H$, toward finding a zero of F by the conventional
Newton's method generally involves finding $h \in H$ such that

$$F'(u)h = -F(u). \tag{25.11}$$

Now u usually needs (when applied to problems in differential equations) to
have some smoothness in order to get a smooth enough solution h so that
when u is updated to $u + h$, the new u has sufficient smoothness. If the
system is of order m, then calculating $F(u)$ from u involves taking m derivatives of u. If u has $k > m$ derivatives, then $F(u)$ generally would have only
$k - m$ derivatives and one would not expect to find h with more derivatives
than this. As the iteration (25.11) progresses, the situation deteriorates to
the point where it can't be continued. This 'loss of derivatives' is countered
by Moser, inspired by Nash's work, [31]. At each step the current value u
is replaced by an approximation to u which has more derivatives. This adds
an inner loop to the process. An eventual proof of convergences is an intricate process requiring a list of additional assumptions on the function F. A
close examination of Problem 401 shows that this iteration, derived from the
continuous Newton's method, does not generally suffer this loss. The key to
avoiding this loss is found in (22.2): The right-hand side of this equation can
be fixed at $-F(x)$ (due to (25.9)) throughout the iteration; all of the linear
systems, essentially

$$F'(y)h = -F(x)$$

for a succession of elements y have the same right-hand side. See [48], [49],
[51], [43] (Chapter 8) for further explanation.

Alfonso Castro, some years ago pointed out to me a relationship between Moser's, [30], inequalities for his inverse function theorem and some of my thoughts on gradients inequalities in Chapter 14. This suggestion resulted in [10] and eventually to the Nash-Moser type results in Chapter 22.

Problem 338 concerns the continuous Newton's method and might well belong to the present chapter.

25.22 Notes on Chapter 23

Before Chapter 23 all semigroups considered were deterministic in the sense that they had the forward uniqueness property, i.e., for a given semigroup T on a space X, if one knows $T(t)x$ for some $t \geq 0$, $x \in X$, then $T(s)x$ is completely determined for all $s > t$ (a slight modification of this statement might be made for local semigroups). In Chapter 23 this condition is relaxed. For p a nonconstant complex polynomial, we interpret the continuous Newton's method as that of finding a continuous function $z : R \to C$ so that

$$z(t_0) = x \in C, \; p(z)'(t) = -p(z(t)), \; t \in R. \tag{25.12}$$

For some $x \in C$, this problem has multiple solutions. This occurs when for some $y \in C$, $s \in R$,

$$p'(z(s)) = 0, z'(s) \text{ doesn't exist and} \lim_{t \to s} z(t) = y. \tag{25.13}$$

It turns out then there are at least two ways to continue $z(t)$ continuously for $t > s$. Such a thing could never happen if (25.12) had a unique solution. Nevertheless, (25.12) does lead to a generalized kind of semigroup, closely following axioms given in [3].

Problem 446 *Do you agree that it is reasonable to call* (25.12) *an equation for the continuous Newton's method?*

The Mathematica code given in Chapter 23 is intended for use in plotting vector fields generated by (25.12). Such plots are in marked contrast to plots obtained with the conventional Newton's method. There are famous plots for complex polynomials arising from the conventional Newton's method. For example, if

$$p(w) = w^3 - 1, \; w \in C,$$

and one colors the complex plane using red, green, blue and black according to the following:

- Color a point red if starting with it, the conventional Newton's method converges to the first root of p.
- Color a point green if starting with it, the conventional Newton's method converges to the second root of p.

- Color a point blue if starting with it, the conventional Newton's method converges to the third root of p.
- Color a point black if starting with it, the conventional Newton's method does not converge.

Similar plots for a Möbius transformation and for other rational functions are of interest. One gets a great fractal mixture of red, green and blue (black ones exist but are not seen in a plot). Such pictures are great for coffee table books, but they are an analyst's nightmare. I consider the 'chaos' represented by the fractal nature of the plot to be an artifact of the discretization of (25.12). Essentially the chaos seen comes from truncation error and is in a sense not natural to the problem.

Problem 447 *References in [44] point to an extension of the above results to polynomials on higher dimensional spaces. Read these references and contemplate an extension of developments in the present chapter.*

Vector fields coming from the Riemann Zeta function suggest a more qualitative approach to the Riemann hypothesis. An idea is this:

In an examination of vector field plots, using the Mathematica code in this chapter but with the polynomial definition replaced by 'Zeta', one notes a quite regular arrangement of arrows in the vector plot. For this I suggest a window, say $\{x, -2, 12\}, \{y, 0, 100\}$, maybe done in pieces which are patched together. As one progresses upward from the real axis, staying fairly close to the critical line, the vector field patterns become a bit more complicated, forming discernible groups, but still quite regular. Particular attention might be paid to the roots of the derivative of the Zeta function. These are points where it appears, in the corresponding vector field plot, that two constant argument lines collide and two leave. Take your choice of which path to follow after such a collision. Now if the Riemann hypothesis is not true, it seems likely that vector field patterns would be severely altered around such a pair of exceptional roots of Zeta. Such a possibility might allow a more topological, global approach to the Riemann hypothesis problem, trying to show that such exceptions can't occur.

25.23 Notes on Chapter 24

As a graduate student in the mid-1950s, in a seminar course under H. S. Wall (all of his courses were like that), we were studying linear evolution equations, essentially problems such as finding

$$u : [a, b] \to X$$

so that

$$u'(t) = A(t)u(t) \ t \in [a, b], \tag{25.14}$$

where X is a Banach space and

$$[a, b] \subset R, A : [a, b] \to L(X, X) \text{ is continuous.}$$

Commonly, for each $t \in [a, b]$, $A(t)$ is a generator of a strongly continuous semigroup and results in semigroup theory are applied to evolution equations.

I asked myself, 'why does everything have to be linear?' A result of this query was [32], my thesis, which dealt with problems of finding

$$Y : [a, b] \to X$$

such that

$$Y(t) = c + \int_a^t dF \, Y, \ t \in [a, b], \tag{25.15}$$

where F is a given function with domain $[a, b]$ and range a set of *nonlinear* functions from X to X. I don't mean that just F itself is nonlinear (this could easily happen in (25.14)) but rather that $F(t)$ itself is a possibly nonlinear transformation for each t in the domain of F. It was at this point that I realized that poor notation from calculus had a bad effect on the development of nonlinear functional analysis. Someone stuck with saying things like

$$f = f(x)$$

was rather frozen out of the subject. It could easily be that f is a function, $x \in D(f)$ and $f(x)$ is also a function, but decidedly $f \neq f(x)$. I had the good fortune of having a rare calculus class, in 1952-53 under R. L. Moore, which maintained good functional notation. This was crucial for me when I began to think about nonlinear evolution equations.

In (25.15), I thought of F as a kind of nonlinear measure-valued function and the integral in (25.15) emulating a Stieltjes integral.

I went from [32] in 1958, to [34] in 1965 (work done much earlier but was slow to be published), in which I started with a collection of functions and derived a function F as in (25.15). Then I wrote [35], published in 1966, in which I found how to approximate resolvents of nonlinear transformations.

Essentially for a strongly continuous nonexpansive semigroup T on a Hilbert space X and for $\lambda, \delta > 0$, I defined

$$A_\delta = \frac{1}{\delta}(T(\delta) - I)$$

and then thought about resolvents:

$$(I - \lambda A_\delta)^{-1},$$

seeking to find when

$$\lim_{\delta \to 0+} (I - \lambda A_\delta)^{-1} x$$

might exist for $x \in X$.

In my 1966 paper, [35], there is a rudimentary exponential formula for a nonlinear semigroup which was under, according to Brezis in 1973, [9], 'hypothèses très restrictives'. I had assumed some differentiability which I wanted to be able to prove (and was later proved by others—see [9], [60],[29]). I agree with Brezis' judgment, but point out that such criticism appeared only after a number of years of work had given improved understanding. At the time of [35] some told me that 'nonlinear functional analysis', was a contradiction in terms. Ten years later it was considered 'mainstream' (whatever that is supposed to mean).

The problems of this chapter use freely ideas from [60] for the development of a form of the Crandall–Liggett Theorem [12] (which in its general form holds in any Banach space). It also uses [9]. These references together with references in [68], [29] contain a great deal for additional study in the direction of this chapter. Other references on monotone operators are [74] and [28] for some very important early ideas on the subject.

After [35], for about 8 years, there was a great deal of activity on nonlinear semigroups, attempting to develop a theory that generalized existing linear theory. A paradigm of that era was to attempt to analyze nonlinear semigroups in a manner *analogous* to the linear theory. Starting in [38], there was an attempt to develop a different idea of generator, one suggested by Sophus Lie's work in which a semigroup was not directly differentiated at zero (to get a conventional generator), but rather to analyze the effect that a semigroup has on real-valued functions when *composed* with a trajectory of a semigroup. This activity, after about 20 years, yielded [14], in which a rather satisfactory theory was obtained linking the set of all jointly continuous semigroups on a Polish space with a clearly defined set of generators in the Sophus Lie sense (Chapters 17, 19).

A quote from Lie's work [27], near the start of his Chapter 4:

"··· **es wird sich nämlich später immer zeigen, dass alle auf die eingliedrige Gruppe bezüglichen Probleme durch Benutzung der infinitesimalen Transformation derselben allein gelöst werden können.**"

A translation coming from an anonymous translator at Oak Ridge National Laboratory reads:

"··· **it will be clear later that all problems related to the one-parameter group may be solved by use of the infinitesimal transformation.**"

In [55], there is a discussion of the ideas that led Gauss and Riemann to their analysis of finite-dimensional surfaces that were *not* specified as a subset of a Euclidean space. This led to Riemannian geometry, where tangent spaces to a manifold are defined as a collection of tangent vectors, each of which is essentially defined as a means of differentiating real-valued functions on the manifold. Our present Lie's generators were adapted from this idea and Lie's ideas underlie the theory in Chapter 17. The work in Chapters 17 and 19 uses only metric spaces without a differential structure; it is essentially a melding of linear semigroup theory (in which generators are only closed and densely defined) with fundamental ideas of Gauss–Riemann–Lie.

Problem 448 *In Chapter 17, if the metric space X is required to be a Riemannian manifold that is not necessarily locally compact (Hilbert manifold), can the theory presented there be enhanced by using a differential structure somewhat similar to what is used for a Riemannian manifold?*

References

1. R. A. Adams, *Sobolev Spaces*, Academic Press, 1978.
2. P. Auscher, S. Hofman, M. Lacey, J. Lewis, A. McIntosh, and P. Tchamitchian, *The solution of the Kato square root problem for second order elliptic operators on R^n*, Ann. Math., 156 (2002), 633-654.
3. J. M. Ball, *Continuity properties and global attractors of generalized semiflows and the Navier-Stokes equations*, Nonlinear Sci., 3 (1995), 475-502.
4. A. Beurling, *Collected Works of Arne Beurling*, edited by L. Carleson, P. Malliavin, J. Neuberger and J. Wermer, Volume I, Birkhäuser, 1989.
5. A. Beurling, *Collected Works of Arne Beurling*, edited by L. Carleson, P. Malliavin, J. Neuberger and J. Wermer, Volume II, Birkhäuser, 1989.
6. A. Beurling and J. Deny, *Dirichlet spaces*, Proc. Natl. Acad. Sci., 45 (1959). USA, 208-215.
7. A. Beurling, *On analytic extension of semigroups of operators*, J. Funct. Anal., 6 (1970), 387-400.
8. K. Brakke, *The surface evolver*, Exp. Math., 1 (1992), 141-165.
9. H. Brezis, *Operateurs maximaux monotones*, North–Holland, 1973.
10. A. Castro and J.W. Neuberger, *An inverse function theorem*, Contemp. Math., 221 (1998), 127-132.
11. M. Certain, *One-parameter semigroups holomorphic away from zero*, Trans. Am. Math. Soc., 187 (1974), 377-389.
12. M. Crandall and T. Liggett, *Generation of nonlinear semigroups of nonlinear transformations on general Banach spaces*, Am. J. Math., 93 (1971), 265-298.
13. J.R. Dorroh and J.W. Neuberger, *Lie generators for semigroups of transformations on a Polish space*, Electronic J. Differential Equations, 1 (1993), addendum attached 1994.
14. J.R. Dorroh and J.W. Neuberger, *A theory of strongly continuous semigroups in terms of Lie generators*, J. Funct. Anal., 136 (1996), 114-126.
15. J.R. Dorroh and J.W. Neuberger, *Linear extensions of nonlinear semigroups*, Progress in Nonlinear Differential Equations, 42, Birkhäuser, 2000.
16. E.B. Dynkin, *Markov Processes-I*, Grund. Math. Wiss., 121, Springer, 1965.
17. K.-J. Engel and R. Nagel, *One-Parameter Semigroups for Linear Evolution Equations*, Springer, 1999.
18. C.L. Fefferman, *claymath.org*.
19. J. Goldstein, *Semigroups of Linear Operators and Applications*, Oxford, 1985.
20. E. Hille, *Functional Analysis and Semigroups*, American Mathematical Society, 1948.
21. E. Hille and R. Phillips, *Functional Analysis and Semigroups*, American Mathematical Society, 1957.

J.W. Neuberger, *A Sequence of Problems on Semigroups*,
Problem Books in Mathematics, DOI 10.1007/978-1-4614-0430-9,
© Springer Science+Business Media, LLC 2011

22. T. Kato, *A characterization of holomorphic semigroups*, Proc. Am. Math. Soc., 25, (1970), 495-498.

23. T. Kato, *Trotter's product formula for some nonlinear semigroups*, Nonlinear Evolution Equations (Proc. Symp., Univ. Wisconsin, Madison, 1977), pp. 155-162, Publ. Math. Res. Center Univ. Wisconsin, 40, Academic Press, 1978.

24. T. Kato, *Abstract evolution equations, linear and quasilinear, revisited*, Functional Analysis and Related Topics, 1991 (Kyoto), 103-125, Lecture Notes Math., 1540, Springer, 1993.

25. P. Kazemi and J. W. Neuberger, *Potential theory and applications to a constructive method for finding critical points of Ginzburg-Landau type equations*, J. Nonlinear Anal., 69 (2008), 925-930.

26. D.G. Kendall, *Some recent developments in the theory of denumerable Markov processes*, Trans. Fourth Prague Conference on Information Theory, Statistical Decision Functions, Random Processes, Academia Prague (1967), 11-27.

27. Sophus Lie, *Differential-Gleichungen*, AMS Chelsea Publishing, 1967 (originally published, in Leipzig, 1891).

28. G. Minty, *On the maximal domain of a monotone function*, Mich. Math. J., 3 (1961), 135-137.

29. R. Martin, *Nonlinear Differential Equations in a Banach Space*, John Wiley and Sons, 1976.

30. J. Moser, *A rapidly convergent iteration method and nonlinear partial differential equations*, Ann. Sc. Norm. Super, Pisa, 20 (1966), 265-315.

31. J. Nash, *The imbedding problem for Riemannian manifolds*, Ann. Math. (2), 63 (1956), 20-63.

32. J. W. Neuberger, *Continuous products and nonlinear integral equations*, Pac. J. Math., 8 (1958), 529-549.

33. J. W. Neuberger, *A quasi-analyticity condition in terms of finite differences*, Proc. London Math. Soc., (2) (1964), 245-259.

34. J. W. Neuberger, *A generator for a set of functions*, Ill. J. Math., 9 (1965), 31-39.

35. J. W. Neuberger, *An exponential formula for one-parameter semigroups of nonlinear transformations*, J. Math. Soc. Japan, 18 (1966), 154-157.

36. J. W. Neuberger, *Lie generators for strongly continuous equi-uniformly continuous one parameter semigroups on a metric space*, Ind. Univ. Math. J., 21 (1971), 961-971.

37. J. W. Neuberger, *Quasianalyticity and semigroups*, Bull. Am. Math. Soc., 78 (1972), 909-922.

38. J. W. Neuberger, *Lie generators for one parameter semigroups of transformations*, J. Reine Angew. Math., 258 (1973), 133-136.

39. J. W. Neuberger, *Nonlinear Semigroups*, Functional Analysis III, Matematisk Inst. Aarhus Univ, 1989.

40. J. W. Neuberger, *Beurling's analyticity theorem*, Math. Intelligencer, 15 (1993), 34-38.

41. J. W. Neuberger, *Predictability in the absence of chaos*, J. Math. Anal. Appl., 175 (1993), 321-332.

42. J. W. Neuberger, *Prevalence of chaotic differences for unpredictable functions*, Acta Sci. Math., 61 (1995), 181-196.

43. J. W. Neuberger, *Sobolev Gradients and Differential Equations*, Springer Lecture Notes in Mathematics 1670 (Second Edition, 2010) (First Edition appeared in 1997).

44. J. W. Neuberger, *Continuous Newton's method for polynomials*, Math. Intelligencer, 27 (1999), 18-23.

45. J. W. Neuberger, *Laplacians and Sobolev gradients*, Proc. Am. Math. Soc., 128 (2000), 853-855.

46. J.W. Neuberger, *Una Sucesión de Problemas de Semigrupos*, Facultad de Ciencias de la Universidad de los Andes, 2000.

47. J. W. Neuberger, *A complete theory for jointly continuous nonlinear semigroups on a complete separable metric space*, J. Applicable Anal., 78 (2001), 223-231.

48. J. W. Neuberger, *A near minimal hypothesis Nash-Moser Theorem*, Int. J. Pure Appl. Math., 4 (2003), 269-280.

49. J. W. Neuberger, *Prospects for a central theory of partial differential equations*, Math. Intelligencer, 27 (2005), 47-55.

50. J. W. Neuberger, *The divergence-free Jacobian Conjecture in dimension two*, Rocky Mountain J., 36 (2006), 265-271.

51. J. W. Neuberger, *The continuous Newton's method, inverse functions and Nash-Moser*, Am. Math. Monthly, 114 (2007).

52. J. W. Neuberger, *Semidynamical systems and Hilbert's fifth problem*, Math. Intelligencer, 30 (2008), 37-41.

53. J. W. Neuberger, *The Tricomi equation: A case study for Sobolev gradients*, Commun. Appl. Nonlinear Anal., (2008), 1-8.

54. J. W. Neuberger, *Lie generators for local semigroups*, Contemp. Math. 513 (2010).

55. D. O'Shea, *The Poincaré Conjecture*, Walker and Company, 2007.

56. G. E. Parker, *A class of nonlinear semigroups with differentiable approximating semigroups*, Proc. Am. Math. Soc., 66 (1977), 33-37.

57. G. E. Parker, *Semigroups of continuous transformations and generating inverse limit sequences*, Pac. J. Math., 80 (1979), 227-235.

58. G. E. Parker, *Semigroup structure underlying evolutions*, Int. J. Math. Math. Sci., 5 (1982), 31-40.

59. A. Pazy, *Semigroups of linear operators and applications to partial differential equations*, Appl. Math. Sci., 44, Springer, 1983.

60. G. da Prato, *Applications croissantes et èquations d'évolutions dans les espacè de Banach*, Academic Press, 1976.

61. N. Raza, S. Sial, S. Siddiqi, and T. Lookman, *Energy minimization related to the nonlinear Schrödinger equation*, J. Comput. Phys., 228 (2009).

62. C. Reed, *The addition of dynamical systems*, Math. Syst. Theory, 6 (1972), 210-220.

63. C. Reed, *Continuous operators that generate many flows*, Dynamical Systems (Proc. Int. Symp., Brown Univ., Providence, R.I., 1974), Vol. II, pp. 207-210, Academic Press, 1976.

64. R.J. Renka and J.W. Neuberger, *Minimal surfaces and Sobolev gradients*, SIAM J. Sci. Comput., 16 (1995), 1412-1427.

65. R.J. Renka and J.W. Neuberger, *Sobolev gradients and the Ginzburg-Landau equations*, SIAM J. Sci. Comput., 20 (1998), 582-590.

66. F. Riesz and B. Sz.-Nagy, *Functional Analysis*, Ungar, 1965 (Dover, 1990).

67. F.D. Sentilles, *Bounded continuous functions on a completely regular space*, Trans. Am. Math. Soc., 168 (1972), 311-336.

68. R.E. Showalter, *Monotone operators in Banach spaces and nonlinear partial differential equations*, Am. Math. Soc. Math. Surv. Monogr., 49, 1996.

69. S. Sial, J. W. Neuberger, T. Lookman, and A. Saxena, *Energy minimization using Sobolev gradients: application to phase separation and ordering*, J. Comput. Phys., 189 (2003), 88-97.

70. J. von Neumann, *Functional Operators II*, Anal. Math. Stud., 22, 1940.

71. J. von Neumann and B. Koopman, *Dynamical Systems of Continuous Spectra*, Proc. Natl. Acad. Sci. USA, 18 (1932), 255-263.

72. A. van den Essen, *Polynomial automorphisms and the Jacobian conjecture*, Sémin. Congr. 2, Soc. Math. France, Paris, 1997, 55-81.

73. G.F. Webb, *Representation of semigroups of nonlinear nonexpansive transformations in Banach spaces*, J. Math. Mech., 19 (1969/70), 159-170.

74. E. Zarantonello, *Solving functional equations by contractive averaging*, Math. Research Center Report 160, Madison, Wisconsin (1960).

Index